Contractor Safety Management

Contractor Safety Management

Edited by
Gregory William Smith

CRC Press is an imprint of the
Taylor & Francis Group, an **informa** business

CRC Press
Taylor & Francis Group
6000 Broken Sound Parkway NW, Suite 300
Boca Raton, FL 33487-2742

Version Date: 20131023

International Standard Book Number-13: 978-1-4665-5684-3 (Paperback)

Library of Congress Cataloging-in-Publication Data

Contractor safety management / edited by Gregory William Smith.
pages cm
Summary: "This book looks specifically at issues for health and safety management that arise in contracting relationships, bringing together a range of perspectives from different disciplines including legal, health and safety management, operational, contract and procurement management. By sharing lessons from both success and failure to identify critical issues in contractor safety management the book raises awareness of the complexity and importance of contractor safety management and to offer some guidance on how those critical issues might be addressed"-- Provided by publisher.
Includes bibliographical references and index.
ISBN 978-1-4665-5684-3 (pbk.)
1. Industrial safety--Management. 2. Contracting out. 3. Industrial accidents--Prevention. 4. Work environment. I. Smith, Gregory W., editor of compilation.

T55.C6635 2013
658.3'82--dc23 2013037629

Visit the Taylor & Francis Web site at
http://www.taylorandfrancis.com

and the CRC Press Web site at
http://www.crcpress.com

Contents

Introduction

After the publication of my first book (Smith, 2011), my publisher was generous enough to ask me if I had any ideas for another book. At the time, I was dealing with a number of issues to do with contractor safety performance and contractor safety incidents. I was working mainly with principals to help them better understand and manage their obligations for health and safety in a contracting relationship.

In a number of cases, the work involved clients whose experience of safety management generally, and contractor safety management in particular, was formed outside of Australia. They were primarily foreign companies that had come to Western Australia to take advantage of the booming mining and oil and gas industries, but had come with little understanding of either the legal requirements or cultural attitudes that influenced safety and health. These requirements and attitudes affected not only the laws and attitudes of regulators, but also the way that principals and contractors worked (or did not work well) together to manage health and safety risks.

Although the lack of understanding was disconcerting, it was also unsurprising. I had been involved in numerous cases where Australian companies had moved into projects overseas with equally limited understanding of their rights and responsibilities for safety and health in different countries, and equally limited strategies for managing health and safety in contracting environments. Indeed, strategies, such as they were, seemed to consist mainly of doing exactly what they had been doing in Australia and assuming that it would be sufficient in the new jurisdiction. What made this strategy all the more surprising is that it did not work particularly well within Australia so there was no real reason to expect that it would work any better anywhere else.

By and large the strategies of large companies coming into Western Australia appear to be the same; do what we did previously and elsewhere, and assume that it will deliver an acceptable result.

Increasingly, companies seemed disappointed by the results.

This book commenced with a general notion of trying to draw together ideas about contractor safety management to distil some key concepts and build guidance for principals and contractors about how to better manage health and safety in contracting relationships. As I began to develop the ideas for the book a number of things became apparent.

First, there seemed to be very few comprehensive publications looking at the issue of contractor safety management. What I could find through my research was the occasional article about contractor safety management or a section about contractor safety management in a broader publication about safety management generally. In other words, books about safety management would include some observations about contractor safety management, but overwhelmingly it was treated as a subset of safety management—to be dealt with for all intents and purposes in the same way as safety management generally.

Second, contracting relationships seemed to be increasingly implicated in major accident events around the world—in recent times, the Pike River Coal Mine* disaster, the Deepwater Horizon disaster in the Gulf of Mexico† and the Montara incident in Australia‡.

In one sense, this is probably unsurprising because major enterprises have always relied on contractors in the conduct of their business. But it did seem to me that there was an increased focus in the relevant inquiries on the relationship between principals and contractors and how that relationship might affect safety performance.

Third, and perhaps running in parallel with the 'involvement' of contractors in major accident events, was the almost universal experience of increasing numbers of contractor 'incidents' in injury statistics. Again, there is probably nothing particularly surprising about that. The changing nature of workforces and working relationships in many countries means that the use of contractors as opposed to employees is more common. I think that it is also true that historically (and currently) contractors are often engaged to do high-risk work.

Anecdotally, however, a number of contractors have suggested to me that they get better safety performance in different contracting relationships, so there may be something in the contracting relationship that influences safety outcomes. I suspect this is an area in need of further research.

Fourth, many clients and businesses that I worked with were devoting more and more time and effort to concerns over their contractor safety performance without, it seemed, much in the way of improvement.

Finally, consistent with my first point—limited material dealing with contractor safety management—the strategies and theories for improving contractor safety management seemed to be simple variations on the

* Underground coal mine explosion in New Zealand on 19 November 2010.

† Drill rig fire and explosion in the Gulf of Mexico on 20 April 2010.

‡ Uncontrolled release of hydrocarbons from a drill rig off the coast of North West Western Australia in August 2009.

themes of general safety management. Without wanting to overuse the term, again this seems unsurprising.

After all, contractors are just another workforce—an employee/employer relationship one step removed—so why wouldn't safety management strategies used in an employment relationship work just as well when I impose them on my contractors?

I think that there are a few observations that we can make here. At a very basic level, for many principals the application of safety management processes in their own businesses are not especially successful so it should not come as any great surprise that imposing those same processes on a contractor would be equally unsuccessful.

Over and above this, however, there are other elements of a principal/contractor relationship that, in my view, influence both safety outcomes and the extent to which a principal can influence those outcomes.

First and foremost is the general short-term, transient and insecure basis of most contracting relationships. Even in relatively long-term contracts, say, 3 to 5 years, in a major project or construction environment 'short-termism' can be a factor. Often a long-term contract moves through a series of stages that requires specialist work groups and additional subcontractors for short periods to manage 'specialised' stages. So even when the 'head contract' and main contractor might be long term the 'workforce' might still be very short term and transient.

In that type of environment safety management directed towards building cultures, resilience and different skillsets may have very limited and even negative effects on safety outcomes. They simply may have no time to take effect, and in the interim could undermine safety efforts.

Although anecdotal, the types of concerns that are often described to me about a principal's safety requirements in a contracting environment include:

- They take up too much time when there is no allowance for that time in the contract. This in turn puts more pressure on the contractor's performance and schedule.
- They create confusion. From the contractor's perspective they have been engaged because they are the 'experts' in performing the work and they have developed processes to manage the risks associated with that work. When a principal imposes themselves on the contractor and starts telling the contractor how to do the work and what procedures and forms they need to use, it simply confuses all parties about how the work should be performed and makes it likely that errors will occur.
- They are reactive and 'over the top'. Most contractors enter into the contracting relationship with an agreed contract and an agreed health and safety management plan to manage the risks of the work

> they are doing. Until there is an incident. Following a safety incident the contractor often finds that they are swamped with a deluge of demands in terms of the incident itself, and new measures that the principal imposes on the contractor. These new measures are often imposed without any regard for the requirements of the contract or the existing health and safety management plan.

The resultant combination of factors is simply a variation on what the Baker Panel Review referred to as 'initiative overload' (Baker et al., 2007, p. 86).

The Baker Panel review was set up to look at BPs safety performance following the Texas City Refinery explosion in 2005. Amongst its many findings, the review identified:

> BP's corporate organization has mandated numerous initiatives to its businesses, including its U.S. refineries, during the last several years. … Each successive initiative has required the refineries to develop plans, conduct analyses, commit resources, and report on progress.
>
> While each initiative has been well intentioned, collectively they have overloaded refinery management and staff. BP's corporate organization has provided the refineries with little guidance on how to prioritize these many initiatives, and the refineries do not receive additional funding to implement each initiative. As a result, senior refinery managers used phrases such as "initiative overload," "incoming," and "unfunded mandates" to describe what they perceived as an avalanche of programs and endeavors that compete for funding and attention. … Many of the hourly workers interviewed at all of the refineries complained that the large number of initiatives and related paperwork contributed to a heavy workload and prevented the workforce from being as focused on safety and operations as they would like. They also reported that the repeated launch of each successive initiative made it increasingly difficult for the workforce to take any of these initiatives seriously; many interviewees described this as the "flavor of the month" phenomenon.

This description captures the long-held sentiment of many contractors that they are endlessly juggling competing and often contradictory demands made by the principal in the name of safety and health. In the end, we are left with a contractor and their workforce overwhelmed with initiatives and safety demands with no time or resources to implement them and no clear guidance or expectations on their priority or how they should be implemented.

Instead of managing the work in accordance with the contract and the agreed health and safety management plan, contractors find themselves having to cope with a moveable ever-changing expectation about the way that health and safety is supposed to be managed.

One of the worst aberrations I have seen of this phenomenon came about on a large 3-year construction project. The project had just over 12 months to run and safety performance had not met the expectations of the principal. At this point, the principal directed the contractor to engage a consultant to drive a safety culture programme in the contractor's workforce. For the life of me, I could not see how the initiative was going to do anything but undermine safety on the project. The thought that we could engage a third party (at the direction of another third party) to 'improve' safety culture in a contractor seemed to me (both at the time and now) to fundamentally misunderstand both safety culture and the short-term nature of the difficulties that the project was facing.

What appears to be emerging is that contractor safety management might be different from just managing safety. That a principal's role in 'overseeing' the safety performance of its contractors is different from managing safety in its own organisation. But, our experience would tend to suggest that by and large contractor safety management is seen as nothing more than a subset of general safety management; that no special consideration needs to be given to understanding the difficulties of the contract environment.

In their excellent book *Resilience Engineering* (Hollnagel et al., 2008), the editors describe their objectives as an opportunity for experts to meet and debate the present state of Resilience Engineering.

Without perhaps attempting anything so ambitious in the content of Contracting this book is at least an attempt to start a conversation about contractor safety management and to promote a debate about what is needed to effectively control health and safety risks in a contracting environment. The book does not represent the end of a journey. Realistically, it may not even be a first step. Perhaps it is something more akin to thinking about packing a bag, knowing that the journey is ahead of us.

In 'packing this bag' I am especially grateful to all of the contributing authors and take this opportunity not just to thank them, but to apologise for my efforts as a first-time editor. What you have before you exists despite my efforts as an editor not because of them.

Finally, can I say that I think that the contractor safety management journey is one worth taking. Not just because of what it can do for the management of health and safety risks in contracting relationships, but what it can add to our understanding of and implementation of safety management strategies more generally.

I invite the health and safety, contract management and risk management community more broadly to think on the topic of contractor safety management. If any readers would like to contribute to future editions looking at contractor safety management, all contributions will be considered and appreciated.

Greg Smith
Legal Practice Director
STE Safety & Legal

References

Baker, J. et. al. 2007. *The Report of the BP U.S. Refineries Independent Safety Review Panel*. U.S. Chemical Safety and Hazard Investigation Board, Washington.

Hollnagel, E., D. Woods and N. Leverson. 2008. *Resilience Engineering Concepts and Precepts*. England: Ashgok.

Editor page

Greg Smith

Legal practice director at STE Safety & Legal

Greg Smith is the legal practice director for STE Safety & Legal. He has spent almost two decades specialising in safety and health management, focussing on assisting clients to understand organisational and individual responsibility for safety and health, and implementing and verifying processes to discharge those responsibilities.

Greg is a qualified lawyer who has worked as a partner and legal practitioner in some of Australia's leading law firms. In addition to his legal experience, Greg has worked as the principal safety advisor for Woodside Energy Limited. In that role he reported to the Vice President–Safety and Health, and was responsible for the ongoing development and implementation of Woodside's global safety management strategy.

Key strategic responsibilities included contractor safety management, incident investigation, the introduction of an organisational behavioural framework to drive cultural change, training and competencies, and developing corporate-level standards and procedures.

Greg is qualified in a range of incident investigation techniques, including TapRoot® and Kelvin TOP-SET®, as well as human factors analysis. Greg is also a qualified lead auditor specialising in health and safety management systems.

As a leading safety and health practitioner, Greg's technical expertise is deep, providing some of Australia's largest and most significant employer's with strategic safety and health advice on compliance, incident investigation management and response, contractor safety management and representation in various legal proceedings. His industry experience is broad, applying his safety and health expertise to the mining, oil and gas, construction, telecommunications, banking, manufacturing, defence, local and state government and transport sectors.

Greg has also devised and delivers comprehensive safety and health training programs and is regarded as a leading provider of safety and health training, particularly in the areas of management responsibilities and contractor safety management. He also teaches accident prevention as part of the School of Public Health, Health, Safety and Environment, at Curtin University in Perth, Western Australia.

Greg has appeared in the Supreme Court, District Court and Magistrates Court of Western Australia, the Federal Court, Equal Opportunity Tribunal and the State Administrative Tribunal. He has also been involved in military boards of inquiry into major accidents.

In 2010 Greg acted for a number of parties in the Montara Commission of Inquiry.

Greg is the author of the book, *Management Obligations for Safety and Health* published by CRC Press.

Greg graduated from the University of Western Australia in 1990 with a Bachelor of Jurisprudence and a Bachelor of Laws and is admitted to practice in the Supreme Court of Western Australia and the High Court of Australia.

Contributing authors

Fiona Murfitt, Master of Risk Management, BBus (Economics/ Marketing/Law), Cert Training & Assessment
DuPont Australia Pty Ltd
Business Director, ANZ and Pacific Islands
DuPont Sustainable Solutions

Fiona's career has predominantly focused on working within high-tisk operations, with more than a decade working within the oil and gas industry. She has held positions in Safety (SHE), Contractor Management and Risk Management, before joining DuPont Sustainable Solutions (DSS), which is DuPont's consulting arm that helps companies build safer, more efficient and sustainable operations.

She now leads DSS in Australia, New Zealand, and the Pacific Islands, being responsible for the businesses growth and strategic development. She is a member of the DuPont Australia Executive Leadership, the Asia Pacific Leadership Team for DSS, and Chairs the Asia Pacific committee for Respect for People.

Fiona's passion around safety, performance, and culture change came after a family member—Kevin, was killed at work. Kevin was a contractor working on a shutdown and the impacts to the family and friends have left a deep impression. As such, being able to influence the improvement of an organisations performance in a safe and sustained way holds a deeply personal connection for her.

Fiona has written many articles and papers related to safety, culture and contractor management. She has been an Industry representative for the Self Insurers Association (Vic), PACIA, and APIA and represented Industry on Government Tripartite working groups. Her thesis paper, The Causal Factors of Catastrophes in Industry focused on learning from incidents. Additional to her formal qualifications, she is also a trained AS4801 auditor, is qualified Root Cause analysis techniques facilitator: TRIPOD, TRIPOD BETA, and Y Tree Analysis, a qualified Bow-Tie Facilitator; MBA Leadership Abridged programme (AIM), and is a DuPont Master Facilitator in the DuPont Integrated Approach (a Behavioural, Cognitive and Social Approach to sustained safety and culture change).

Elias M. Choueiri, PhD

Prof. Dr. Choueiri is the author/co-author of 15 books, including the following

- **Choueiri, E.M.,** *Trips Towards Traffic Safety,* Dar Al Ma7aja Al Bayda2, Lebanon, 2010; p. 64, in Arabic.
- **Choueiri, E.M.,** *Guide for Drivers and Pedestrians,* Lebanese Association for Public Safety, Lebanon, 2010; p. 133, in Arabic.
- **Choueiri, E.M.,** *Saleh and His Story with Traffic Safety,* Dar El Ilm Lilmalayin, Lebanon, 2009. p. 56, in Arabic.
- **Choueiri, E.M.,** Nashef, A., and Saade, W., *Traffic Accidents: Between the Experts' Reports and the Court's Judgments, in Accordance with the Traffic Law,* Al Ghazal publisher, Lebanon, 2008; p. 458, in Arabic.
- **Choueiri, E.M.,** Hindi, T., and Akl, Z., *The Citizen's Charter for Public Safety,* Republic of Lebanon, Office of the Minister of State for Administrative Reform, Lebanon, March 2007; p. 72 (English), 76 (French) and 79 (in Arabic).
- **Choueiri, E.M.,** *Guide for Drivers and Pedestrians: Their Rights and Duties,* a Special Annex to Al-Amn Magazine, No. 146, March 2004; 40 pages.
- Lamm, R., Psarianos, B., Mailaender, T., **Choueiri, E.M.,** Heger, R., and Steyer, R., *Highway Design and Traffic Safety Engineering Handbook.* McGraw-Hill, Professional Book Group, New York; p. 1088; Language Editors: Hayward, J.C., **Choueiri, E.M.,** and Quay, J.A., 1999.

He also has over 300 refereed publications, technical reports, conference presentations and newspaper articles.

He pursued his higher education studies at several universities in the United States, which culminated with a Ph.D. in engineering science, with a concentration in transportation engineering, from Clarkson University, Potsdam, New York, in 1987. He holds several graduate and undergraduate degrees in civil and environmental engineering, electrical and computer engineering, mathematics and computer science, and mathematics.

His research interests are mainly in the areas of safety education, safety management, highway design, traffic safety, driving dynamics, driver behaviour, and railway transportation. He has won 19 awards for his scholarship and has held faculty and managerial positions at several universities in the United States and Lebanon.

He serves (and served) on the editorial boards of a number of scientific journals. He is the president of the Lebanese Association for Public Safety (LAPS), Lebanon. He sits on the board of directors of the

World Safety Organization (WSO), chairs the WSO Highway Transport Committee, chairs the WSO Transportation of Dangerous Goods Committee, and serves as WSO Liaison Officer to the United Nations.

Janis Jansz, RN., RM, Dip.Tch., BSc, Grad.Dip. OHS, MPH, PhD, FSIA
Dr Jansz is employed as a senior lecturer, Occupational Health and Safety Environmental Health, Curtin University and has an adjunct senior lecturer appointment at Edith Cowan University in the School of Management. Since 1996 Janis has been the director of the International Labour Organisation (ILO) Communications, Information, Safety (CIS) Centre in Western Australia. She was a member of the executive committee of the Safety Institute of (Western) Australia Inc. from 1990 to 2010 and was the first female president from 1997 to 2000. Janis was editor of the *Australian National Safety Journal* from 1994 to 2000. She was given the Safety Institute of (Western) Australia Inc. Member of the Year Award in 1994 and in 1999 for her professional work in improving occupational safety. Dr Jansz is a member of the Curtin Health Innovation Research Institute, the World Health Organisation Collaborating Centre for Environmental Health Impact Assessment, Centre for Research in Entertainment, Arts, Technology, Education and Communications and a member of the Curtin–Monash Accident Research Centre. Since 1988 she has been a member of the Occupational Health Society and is currently an Executive Committee member of this organisation. Since 1997 she has been Director, World Safety Organisation National Office for Australia, member of the board of directors for the World Safety Organisation and editor of the *World Safety Journal* from 2002. She continues to hold all of these positions. In 2005 and in 2011 Dr Jansz was awarded the World Safety Education Award for her contribution internationally to providing occupational safety and health education. In recognition of her professional work in improving occupational safety and health worldwide through her teaching, research and professional service, Dr Janis Jansz was presented with the award of World Safety Person of the Year at the World Safety Conference in the United States in 2001.

Dr Jansz began her career working as a registered nurse where she cared for people who were injured, ill, and terminally ill people. As a registered nurse and midwife she worked in a variety of city and country nursing positions as a clinical nurse in most areas of nursing and in a range of managerial nursing positions. She has had experience working as an agency nurse in a wide variety of private and government hospitals. Dr Jansz enjoys working as an occupational safety and health professional because she has the opportunity to improve people's health, the work environment, work processes, management and business profitability whilst preventing people from becoming ill, injured or dying due to work-related causes. Dr Jansz appreciates being able to share occupational safety and health knowledge with other people through teaching, research and writing activities. Author of over 100 journal articles, textbook chapters and conference papers, she has written the distance education material for 18 units of occupational safety and health study for two universities. Research and teaching activity is centred on occupational safety and health management, ergonomics, communicable disease control, health promotion, workers compensation and injury management, safety inspections, audits and risk management, accident prevention, and on developing safety management plans, occupational safety and health policies, procedures and programs.

Patrick Gilroy, AM
Chief Executive Officer
Mining and Resource Contractors Safety Training Association
Patrick Gilroy was General Secretary/Deputy Chief Executive Officer, Chamber of Minerals and Energy of Western Australia from 1982 to 1999 with responsibility for the industry OSH portfolio during that period.

In 2002 he was made a member (AM) in the general division of the Order of Australia for his services to occupational health and safety in the mining industry.

He was a member of the WorkSafe Western Australia Commission as an industry representative and latterly as an independent expert from 1987 to 2003.

He was awarded a Centenary Medal in 2003 for services to the Commission.

He is currently the CEO of MARCSTA.

Patrick has made a long-standing and wide-ranging contribution to safety, including:

WorkSafe WA Commission

- Employer Representative (1986–1996)
- Expert Member (1996–)
- Chairman Fatalities Working Party (1998–1999)

Mines Occupational Safety and Health Board

- Member of the Interim Board
- Member of the Statutory Board (1994–2000)
- Member of the Occupational Safety and Health Standing Committee (1994–2000)
- Member of the Prevention of Mining Fatalities Taskforce (1997)
- Member of the Risk Taking Behaviour in the Western Australian Underground Mines Sector (1998)

Major Activities

- Drafting Committee to develop legislation to compensate for Noise Induced Hearing Loss (Workers' Compensation and Rehabilitation Act)
- Drafting Committee for both the Occupational Safety and Health Act and the Mines Safety and Inspection Act.
- NOHSC Advisory Group on the Australian Standard for Respirable Silica.
- Coordinator of Minesafe International (1990, 1993, 1996, 1998, 2000), the world's major conference on occupational health and safety in the mining industry. Proceedings published on all occasions.
- Editor of the *Occupational Health and Safety Bulletin* of the Chamber of Minerals and Energy of Western Australia Inc. (1990–1999)

Olga Klimczak
LLB (Hons.) BA(Hons.)
LLM candidate, University of Melbourne
Olga is a senior associate based in Perth, Western Australia, and works for a leading international firm, specialising in workplace safety and health, employment, industrial relations and diversity. She has extensive experience advising a range of government and private sector clients in a number of industries, including the resources sector (mining, oil and gas), manufacturing, maritime, professional services and banking. Olga has a practical and commercial approach in assisting her clients.

Olga's experience includes advisory work, presenting training and seminars, representation in dispute resolution and court/tribunal proceedings, investigations and audits, and document management.

Olga completed a Bachelor of Laws and a Bachelor of Arts, both with honours, from the University of Western Australia in 2005.

In 2006, Olga worked as an associate to the Honourable Justice McLure of the Western Australian Court of Appeal, who is now the president of the Court of Appeal.

Olga worked in London and travelled in 2007, before commencing her current role in September 2007.

Olga is currently studying her Master's of Employee and Labour Relations Law at the University of Melbourne, which has included units in safety law.

She is admitted to practice in the Supreme Courts of Western Australia and the Federal and High Courts of Australia, and has appeared as counsel in the Western Australian Supreme, District and Magistrates Courts, the Federal Magistrate's Court and Fair Work Australia.

Tristan Casey DPsy(org), BPsycSc
Sentis
Tristan Casey is an experienced and skilled research analyst at Sentis. His primary research interests include safety climate, safety leadership, human factors, training transfer and evaluation, human–computer interaction, and survey-based research methods. Tristan is passionate about synthesising and translating empirical state-of-the-art research into practical applications, such as the development of evidence-based innovative products and services. Tristan also has a particular interest in developing research instruments to identify strengths and diagnose challenges within organisations and teams. At Sentis, Tristan has developed numerous research-based tools and offerings such as the Safety Leadership Assessment and the Safety Climate Survey, and acted as lead researcher on numerous large-scale research projects across oil and gas, mining, and utilities industries. As a co-investigator, Tristan is currently progressing applied research programs in error management and safety training transfer. The objectives of these programs are to develop and evaluate evidence-based tools to increase worker safety and contribute to both practical and empirical understanding of these emerging areas in occupational safety management. Tristan has presented his research at numerous psychology conferences throughout Australia (e.g., International Congress of Applied Psychology, Australian Psychological Society, Industrial and Organisational Psychology) and published his research findings in peer-reviewed journals (e.g., Safety Science and Computers in Human Behavior).

Sarina M. Maneotis, MS
Sentis
Sarina (Sari) Maneotis is a member of the Research Team at Sentis. She supports the analysis and reporting of Sentis' safety climate surveys and safety leadership assessments. In her role, Sari also develops and validates leadership and well-being assessments. Recently, Sari contributed to the development of Sentis' 360-degree executive leadership assessment as well as Sentis' well-being climate survey used to diagnose employees' wellbeing challenges at work. Sari also serves as the lead researcher for intervention efficacy studies conducted in partnership with Sentis' North American clients.

Prior to joining Sentis, Sari was a research team member for a project on creativity and leadership that was funded by a grant from the National Science Foundation. She has also consulted on various employee selection, performance appraisal, and employee survey projects whilst completing her master's degree in Industrial/Organizational Psychology at Pennsylvania State University. In addition to her focus on leadership and wellbeing at Sentis, Sari has conducted research on emotion regulation within the customer service context. Her research has been presented at several professional conferences, and her master's thesis, focusing on pro-social motivation and emotional labour, has been accepted for publication at the journal Human Performance. Sari is also currently completing her dissertation at Pennsylvania State University, which focuses on service employee well-being.

Autumn D. Krauss, PhD
Sentis
Autumn Krauss is Chief Scientist at Sentis, an occupational health and safety consultancy based in Brisbane, Australia. In her role, she manages Sentis' global Research Team who partners with clients to conduct applied research within the domains of employee wellbeing, workplace safety, and leadership and organisational development. Autumn also directs the Sentis Academy, which houses all of Sentis' internship schemes and university partnerships around the world. Her applied research programs have been funded by the Centers for Disease Control and Prevention, the National Institute for Occupational Safety and Health, the Society for Industrial and Organisational Psychology, and the Society for Human Resource Management. She has authored chapters in books such as Contemporary Occupational Health Psychology and International Review of Industrial and organisational Psychology. She has co-authored research articles in journals such as *Accident Analysis and Prevention* and *journal of Business and Psychology*. Autumn has spoken on workplace safety topics in the United States, the United Kingdom, and Australasia for organisations such as the American Society of Safety Engineers, the Institution of Occupational Safety and Health, and the Safety Institute of Australia. She has also presented the results of studies conducted by the Sentis Research Team at the SIOP Annual Conference, APA's Work, Stress, and Health Conference, and the Congress of the European Association of Work and Organisational Psychology. While Autumn's current work is focused on occupational health in high-risk industries such as mining, construction, and energy, she has also consulted on talent management challenges with organisations in industries such as retail and food service. She holds a PhD in Industrial and Organisational Psychology with a specialisation in Occupational Health Psychology from Colorado State University.

chapter one

Contractor safety management

Concepts and issues

Greg Smith, BJuris. LLB
Legal Practice Director
STE Safety & Legal

Contents

Introduction

I am not a golfer. I have played two games of golf in my life, both of which I was compelled (in one case, ordered) to play.

My first game was in 1993 when I was 'ordered' to participate in an officers versus sergeants mess match. Without going into the quality of my effort, I was awarded five 'extra duties' by my commanding officer at the time—extra duties being a form of 'unofficial' disciplinary action for poorly performing junior officers.

My second game was in Thailand in 2008. I was in the country as part of a management team for an oil and gas company involved in some major component construction work. My boss at the time was a mad keen golfer, and although it was not compulsory, there was certainly a sense of expectation about playing. In the end, my boss may have regretted my involvement, and although he did not have the disciplinary option that my former commanding officer had, it did not seem to diminish his obvious desire to mete out something similar.

I have a photograph from that day, which shows a plume of dirt and grass. Behind that, you can see a golf club three quarters of the way through a swing and a glimpse of a green shirt; it is me, teeing off!

However, why open with golf? Well, to me golf has always been a simple game in theory. Hit a small ball with a stick, down a more or less straight line to a hole in the ground and tap it in.

Obviously, the trick is in the execution.

Similarly, safety management theory is quite simple: Identify the hazards, all the things that could cause 'harm' in your business. Assess the risks, or the likelihood that those hazards will cause harm. Develop controls to manage those hazards and associated risks, and ensure that those controls are in place and are effective to manage the hazards and risks.

Again, the trick is in the execution.

In recent years, a number of major accident inquiries have identified that the relationship between the principal and the (typically) multiple levels of contractors engaged in their enterprise has played a contributory role in disasters.

What emerges is that there are aspects of contracting relationships that can create special challenges for safety management, or perhaps add increasing layers of complexity to the challenges that already exist. Just by way of example, some key safety management challenges that take on different complexities in a contracting relationship include

- The efficiency/thoroughness trade-off and
- Incident investigations.

The efficiency/thoroughness trade-off

The efficiency/thoroughness trade-off, sometimes referred to as 'sacrifice decision making' (see, for example, Hollnagel et al., 2008) requires a balance of:

- The steps to be taken to ensure safety

against

- Getting on with doing a job

There is nothing remarkable about this tension, and it represents no more than the ordinary cost, time, resources and schedule pressure that exist in every business.

One framework for considering the efficiency/thoroughness trade-off is a legal framework.

A typical legal framework is that the law does not require a business or employer to prevent all accidents; rather, it imposes boundaries of legally acceptable or defendable behaviour. These boundaries include such notions as 'reasonableness', 'foreseeability', 'due diligence', 'negligence', 'practicable', 'recklessness', 'carelessness' and so on.

If there is an accident and it can be shown that my behaviour as an individual or an organisation is 'outside' of the boundary, then my conduct may be regarded as legally 'blameworthy'. If my conduct is within the boundary, it is not legally blameworthy; it is not unlawful. Therefore, whilst every accident in a workplace may be regarded as a failure of the safety management system, not every failure is unlawful.

Of course, the law is just framework within which the efficiency/thoroughness trade-off can be considered. Other 'frameworks' include:

- *Commercial:* What is the cost/benefit analysis? How much am I prepared to gamble that a health and safety hazard will not manifest itself in accident in order to achieve a better budget, production or schedule outcome?
- *Reputational:* When does the reputational risk (both personal and organisational) attached to a potential accident override strict, immediate legal and/or commercial concerns?
- *Moral:* At what point do my personal standards and 'moral compass' compel me to do more for safety and health in my workplace when it may not be 'legally' required, nor in the best commercial interests of the business?

There is also, of course, an inherent business risk particularly from major disasters. We have seen in recent decades how major health and safety disasters have the potential to significantly undermine the validity of, and in some cases destroy, businesses.

The efficiency thoroughness trade-off is compounded in a principal/contractor—or even more remote, principal/contractor/subcontractor—relationship.

So, for example, if I have engaged expert contractors to do work for me and they are contractually obliged to do the work 'safely', how much time and effort

a. should I be expected to put forth, and
b. do I want to spend in making sure that the contractor is doing their job?

Equally, if I am a contractor engaged for my expertise, how much time should I have to spend dealing with my principal's safety demands—especially if I have signed a contract and provided a health and safety management plan that the principal has approved, which clearly sets out how safety will be managed?

Tension and conflict in the management of safety can often arise during the life of the contract, especially as cost and schedule pressures come to bear on the performance of the contract. As the performance of the contract comes under pressure, this can often lead to an increase in the number of incidents and near-misses. The fallout of this is often greater demand from the principal for actions to be taken on safety, and at that point different factors can start to come into play.

By demanding safety actions over and above what was agreed in the contract, the principal may be adding further to the cost/schedule pressures, and in turn further undermining safety. If the relationship between the parties is strained, it may become increasingly difficult to reach a sensible compromise and/or contract variation that can help balance the safety expectations of the principal with the contractor's ability to deliver against the contract.

Incident investigations

It is almost trite to say, but good incident investigations are a critical factor in any effective safety management system.

There are substantial challenges within an organisation to produce high-quality, meaningful incident investigations that give insight into how well health and safety risks are being managed. In a contracting relationship where safety failure can have important legal and commercial consequences, the level of insight offered by the incident investigation process can become very cloudy indeed.

Typically, a principal is interested in understanding what the contractor did 'wrong'. A contractor is often trying to walk a very fine line between downplaying their level of 'fault' and avoiding being overly critical of the role that the principal might have played in the incident. It seldom enhances the contractor's reputation to document and point out management system failures of their client.

Two recent enquiries into major disasters in the oil and gas industry has criticised the way that the principals sought to downplay their role in the incident or, indeed, failed to adequately consider their role at all (Graham et al., 2011; Borthwick, 2010).

Very often, the investigation of a 'contractor' incident completely overlooks the role of the principal's management system, meaning that whilst individual contractor errors might be identified and addressed, the broader failure is contractor safety management, with the potential to undermine safety much more widely going unchecked.

For most investigations where the incident involves a contractor, the primary (and only) focus of the investigation is to understand the 'cause' of the specific incident, typically, what the contractor did or did not do.

By way of example, one investigation that I reviewed involved an injury to a drilling supervisor employed by a subcontractor. The injury occurred when a drill string, weighing approximately 1.3 t, slipped out of the clamp that was holding it in place and hit the drilling supervisor on his arm. The drilling supervisor suffered quite a serious injury, but the incident could easily have been fatal if he had been positioned only slightly differently.

The evidence in the investigation was quite clear: the drilling supervisor ought to have known better, he had been warned by other members of the drilling crew not to position himself in the 'line of fire', and he had been involved in previous risk assessments that identified 'line of fire' as a risk.

The focus of the investigation centred on technical reasons why the clamp holding the drill string failed, the drilling supervisor's knowledge about what he should and should not have done and, to a lesser extent, 'cultural' elements that may have contributed to the incident, including possible production pressures, cultural performance amongst drillers and so on.

What the incident investigation did not consider at all was the principal's obligations in engaging and managing contractors.

The principal had a detailed contract safety management system that described how contractors would be selected, engaged and managed. In addition, there was a documented contract that described the obligations of various parties, and the contractor had provided a health and safety management plan that described how the contractor and subcontractors would manage the risks on the project. The principal also had a health and safety management plan.

By analysing the various health and safety documents, we were able to identify at least 35 individual elements that ought to have been in place to manage health and safety in the contracting relationship. Those elements include things such as

- Appointing a contract manager
- Issuing permits to work
- Documented workplace inspections
- Inspecting the contractor's equipment before it came on site
- Receiving monthly and weekly health and safety information from the contractor
- Closing out audit non-conformances before being allowed to start work

Ultimately, we could only identify half a dozen of these requirements that had been complied with, and in each case, compliance was below an acceptable level. However, the investigation team had not turned its mind to these issues at all.

The failure to investigate the principal's own contractor safety management system left a significant gap in their understanding of the effectiveness of safety management on the project. At best, the resultant investigations meant that the principal may have been able to address the specific risks arising from the specific incident (including 'line of fire'), or even drilling operations, more broadly. However, they had no understanding of whether or not the contractors had been engaged in accordance with their safety management systems (systems that were designed to ensure safety!), and/or the extent to which the failures of their contractor safety management system extended across the project to other contractors.

The formulaic approach to contractor safety management, like safety management and golf, suggests that the task of contractor safety management should be easy. Conventional wisdom suggests a structured approach consisting of:

1. *Determining the contract requirement:* Wherein the principal determines the health and safety requirements/expectations of the work to be done before it contracts the work out. Best practice would suggest that this also requires an assessment of any risks arising because of the decision to contract workout.
2. *Assess the contractors:* Having gone to market to seek contractors, the principal then assesses the contractor best placed to meet the principal's health and safety requirements and expectation. Of course the 'best' contractor from a health and safety perspective may not be the 'best' contractor. It would be naïve to suggest that cost and availability (i.e., when they can start work) do not play a significant, if not overriding, role in the ultimate selection of the contractor.
3. *Awarding the contract:* From a health and safety perspective this usually means negotiating and agreeing the health and safety terms of the contract and how health and safety will be managed in the performance of the contract works.
4. *Managing the contract:* This refers to the process whereby the principal ensures that the contractor is actually performing the work as agreed and that they are managing the health and safety risks in accordance with the contract.

The Western Australian Supreme Court, discussing the principal/contractor relationship in the context of working at heights, expressed some of these principles in the following way (my emphasis added in bold):

> If there had been evidence about safety meetings being held with [the contractor] concerning safety and concerning the steps to be taken by the sub-contractor concerning safety, then this may have been significant evidence. Such a meeting may have resulted in assurances by the sub-contractor

> to the appellant about steps it would take to ensure the safety of its workers. An item for discussion at such a meeting should have been about who would install anchor points for harnesses to be used by roof workers. That may have resulted in the appellant installing the anchor points or may have resulted in the sub-contractor promising to do so. **If the latter, then the appellant would have been obliged to check to see that the promise had been fulfilled**. If the anchor points had been installed and if the appellant then saw that safety procedures agreed to were being followed, then the fact that on one day when Mr Hughes was not in attendance the safety procedures were not followed, may not have afforded any evidence to sustain the charge.*

Given all of the potential commercial, systems, cultural and priority conflicts that can arise between various entities, it is perhaps unsurprising that many major disasters of the past few years have involved layered contracting relationships.

Recent examples

The nature of these complex relationships, and the impact they can have on effective safety management was the subject of considerable discussion in the recent various enquiries into the events surrounding the Deepwater Horizon disaster in the Gulf of Mexico in April 2010.

That event culminated in a fire and explosion on a deepwater drilling rig (the Deepwater Horizon) in which 11 people died and hydrocarbons continued to flow uncontrolled from the well for more than 80 days, leading to one of the worst oil spills and pollution events in United States history.

BP had purchased the rights to drill for hydrocarbons in the Gulf of Mexico and was the legal operator for the activities needed to complete the drilling operations. However, a range of contractors on BP's behalf undertook the physical work of constructing and drilling the well.

Key contractors included:

Transocean: Transocean was in the business of contracting drilling rigs. Transocean's crews performed most of the basic drilling work, and Transocean employees were in charge of the Deepwater Horizon.

* *Silent Vector Pty Ltd v Shepherd* [2003] WASCA 315 [22].

Halliburton: Halliburton is an oil field services provider who designed and pumped the cement needed for critical work to secure the pipe work (drill string or casing) as it was drilled into the seabed and beyond. Halliburton employees worked closely with BP onshore as well as on the rig.

There were at least six other key contractors and suppliers identified as having important roles to play in the work on the Deepwater Horizon.

The extent to which BP relied on contractors to do the work of drilling the well is emphasised by the fact that on the day of the fire and explosion only 7 of the 126 workers on the Deepwater horizon were BP employees (Bartlit, Sankar and Grimsley, 2011, p. 30).

The presidential inquiry into the Deepwater Horizon disaster (Graham et al., 2011) found (amongst other things) that better communication between BP and its contractors would have played an important role in preventing the incident. Moreover, they said that organisations must have effective systems in place for integrating contractors into high hazard and complex operations such as deepwater drilling. These systems need to consider corporate cultures, internal procedures, and the different decision-making processes of the different entities (Graham et al., 2011, p. 122).

The presidential inquiry went on to say:

> . . . the extensive involvement of those contractors in the mistakes that caused the Macondo well blowout underscores the compelling need for a fundamental shift in industry culture . . .
>
> . . . whatever the specific contractual relationships, operating safely in this environment clearly demands a safety culture that encompasses every element of the extended drilling services, and operating industry. (Graham et al., 2011, p. 223)

Referring to evidence given by the chair of the University of Texas's Department of Petroleum and Geosystems Engineering, Tad Patzek, at an earlier congressional hearing, the presidential Inquiry accepted Patzek's view that:

> . . . individual contractors have different cultures and management structures, leading easily to conflicts of interest, confusion, lack of coordination, and severely slowed decision-making. (Graham et al., 2011, p. 229)

Chapter 5 of the chief counsel's report into the Deepwater Horizon disaster includes a specific section on 'Contractors' (Bartlit, Sankar and Grimsley, 2011, p. 225).

Chapter 5 is titled 'Overarching Failures of Management'.

The chief counsel's report found that the various parties appeared to lose sight of the potential risks in contracting relationships, including risks arising from miscommunication and misunderstanding (Bartlit, Sankar and Grimsley, 2011, p. 238). The report identified three key areas of 'failure' in the principal/contractor relationship (Bartlit, Sankar and Grimsley, 2011, pp. 238–240) that had a role to play in the disaster.

BP's oversight of contractors: The report found that BP did not adequately supervise its contractors in a number of important ways. The main example relied upon to support this position was BP's supervision of cementing work carried out by Halliburton.

Although BP argued they had relied on Halliburton's expertise in relation to cementing, documentary evidence indicated that BP had a number of concerns about their cementing services, and specific concerns about the competence of a particular cementing engineer, Jesse Gagliano. These concerns extended to BP asking Halliburton to reassign Mr Gagliano.

Notwithstanding these concerns, it appears that BP did not take any additional steps to oversight to work undertaken by Mr Gagliano. The chief counsel's report identified that, in light of these concerns, at the very least BP should have ensured there was some 'double-checking' of Mr Gagliano's work.

Contractors' deference to BP: The problems associated with a lack of oversight of a contractor were compounded in the case of the Deepwater Horizon by what the chief counsel found to be the contractors' deference to BP. In one case, a technician engaged by another contractor, Weatherford, described his role as being to do what the company requests.

The chief counsel's report identified numerous instances where contractors had concerns about elements of the work being performed, but did not forcefully pursue these concerns with BP.

It is perhaps worth pausing at this stage to consider that the deference of contractors to their principals is not always the case. In the Montara Commission of Inquiry discussed below (Borthwick, 2010), considerable attention was given to correspondence from a contractor who withdrew its services from the project because it was concerned that decisions were being made by the principal based on cost and schedule, rather than on good engineering (i.e., safety) principles (Duncan, 2010, pp. 1390–1391).

Lack of clarity about contractor expertise and responsibility: Finally, the chief counsel's report identified that there was a lack of clarity, in particular between BP and Transocean, about the competence of Transocean's workers on the rig to interpret data that was critical to the safe performance of the work.

Whilst it is highly likely that much of this confusion arose 'after the event' as each party tried to sheet some of the blame home to the other, what was clear (and is very important in a contractor safety management context) is that there was no real clarity around decision making and decision-making authority based on technical *competence* before the incident.

Less than 12 months prior to the Deepwater Horizon disaster, a strikingly similar event occurred offshore from North West in Western Australia. In August 2009 an uncontrolled release of hydrocarbons occurred from the wellhead platform in the Montara Development Project. In all practical senses, the contractual arrangements in place at the time of the Montara incident were very similar to those that existed on the Deepwater Horizon.

PTTEP Australasia (Ashmore Cartier) Pty Ltd (PTT) was the legal 'operator', but at the time of the incident, operations were being conducted by a third-party entity, West Atlas, which owned the drilling rig and undertook the drilling work. Like the Deepwater Horizon, cementing played a significant role in the Montara incident, and again the cementing contractor was Halliburton.

Without identifying 'contractors' in a specific section of the inquiry report, nevertheless, the Montara Commission of Inquiry did make a number of findings relevant to contractor safety management (Borthwick, 2010, pp. 10–12):

- Records management and communication between the parties were deficient.
- There was a 'systemic failure' of communication between the parties.
- The relationship between PTT and West Atlas (as the rig operator) was deficient, and there were ineffective exchanges of information between them.
- The relationship between PTT and West Atlas needed to 'be more formalised' with explicit 'sign off' between the parties on important decisions that had the potential to affect safety.
- PTT's governance structures, up to and including the chief executive officer and parent company in Thailand, paid insufficient attention to managing the project risks, including (specifically) their 'interaction with contractors'.

Similarly to the Deepwater Horizon incident, the Montara incident was first and foremost an issue of well control. The Montara Commission of Inquiry found it was incumbent on PTT and West Atlas to develop clear protocols for working together, including, importantly, well control operations. At one point, the commission described some of the documented safety management processes as 'replete with delphic motherhood statements' (Borthwick, 2010, p. 135) such as

> Safety management in the field is primarily the responsibility of the Vessel Masters/Superintendents, FPSO OIM, Rig OIM and WHP Person In Charge (PIC). The prioritisation of all activities in the Montara field is the responsibility of the PTTEPAA Project Manager. However, control of the individual activities during the field development remains with the relevant supervisors.
>
> …
>
> All parties in the Montara field development shall have clear structuring of HSE interfaces to ensure that there is no confusion as to: approval authority; roles and responsibilities of personnel; organisational structures, management of HSE; operating procedures; reporting structures; and SIMOPS.

It is not overly difficult to ascribe actual responsibility for doing anything to anyone or no one under the terms described above.

The commission identified that deficiencies in respect of planning and clarity of roles and responsibilities constituted *'one of the most significant indirect causes of the Blowout'* (Borthwick, 2010, p. 133).

Whilst these cases consider contractor safety management in the context of the interaction of the principal and the contractor and how that interaction (or lack of it) might contribute to the incident, there are other aspects to contractor safety management.

A good example of this is the BP Texas City Refinery explosion (see, for example, CSB, 2007 and Baker et al., 2007).

On 23 March 2005, one of the worst industrial accidents in the history of the United States occurred when the BP refinery at Texas City exploded during start-up operations. The disaster resulted in 15 deaths and more than 170 injuries.

The U.S. Chemical Safety and Hazard Investigation Board (CSB), an independent federal agency charged with investigating industrial chemical accidents, undertook an accident investigation into the incident. The investigation ultimately found that a string of technical, procedural, leadership, management and safety culture deficiencies combined to cause the incident (CSB, 2007, p. 25).

As part of their investigation, the CSB recommended that BP commission an independent panel to assess and report on the effectiveness of BP's oversight of safety management and safety culture of its refineries in North America. In October 2005, BP announced the formation of the BP U.S. Refineries Independent Safety Review Panel, to be chaired by former Secretary of State James A. Baker III (Baker et al., 2007).

However, in that case most of the workers killed were contractors who had nothing to do with the work that was being performed and that led to the explosion.* These workers, and others, were using temporary office accommodations—trailers—that had been positioned too close to the hydrocarbons processing unit that was being started. It was errors that were made during the start-up process that caused the explosion. These trailers had been positioned in breach of BP's own risk management requirements.

This creates a different scenario from a number of the examples above; this case demonstrates not just the importance of ensuring that contractors are competent and working safely, but also that the contractors are 'protected' from the work of the principals', employees, as well as work being performed by other contractors.

Earlier I touched on concepts of frameworks within which organisations might consider the efficiency/thoroughness trade-off. These frameworks become even more convoluted once multiple contractors or multiple layers of contractors become involved. If we simply consider the legal framework, typically a key component of a contracting arrangement is to transfer legal risk through mechanisms such as insurance provisions, indemnities and similar hold harmless clauses.

The question that arises, legitimately, is how much time and effort do I want to spend monitoring a contractor's safety performance if I believe that they are carrying all of the legal risk under the contract? The answer to that question might depend on a number of other efficiency/thoroughness trade-offs that we have touched on, or indeed the consequences of an incident.

Consider for a moment the Deepwater Horizon disaster in the Gulf of Mexico discussed previously. If we assume for a moment that BP's contractual arrangements with Transocean were watertight and that all legal and commercial risk was able to be sheeted home to Transocean, presumably that still would have been an accident worth preventing for any number of reasons unrelated to legal liability and commercial loss.

The management of safety and health is a basic obligation in any working environment. For a safety purist, this means that the health and safety risks just need to be controlled irrespective of who is 'legally', 'contractually' or 'commercially' responsible.

* Some 11 workers who died were employees of J.E. Merit, part of Jacobs Engineering in Pasadena, California. *Houston Chronicle* 24 March 2005. http://www.chron.com/news/article/15th-body-pulled-from-rubble-of-BP-s-Texas-City-1936338.php.

Business reality is that using contractors is a commercial necessity that by implication carries with it consequences of risk transfer. By design, contracts are structured to try to pass risks between the parties.

Further, in many cases the only responsible (as well is legal) way to manage health and safety risks is to engage a specialist contractor with the skills and expertise to undertake work that the principal does not have skills or expertise in. In this environment, we simply cannot manage health and safety risks regardless of who is responsible. That would just put safety management out of step with normal business objectives and risk marginalising safety from the business rather than integrating into it. Moreover, it makes the workplace unsafe. If there is no clear, and clearly understood, assignment of rights and responsibilities for all aspects of safety management in a contracting environment, it is likely that no one will be 'responsible' and safety will fall through the cracks. We will have failed in our fundamental obligation to know that our health and safety risks are being controlled.

As the chief counsel's report (Bartlit et al. 2011, p. 227) into the Deepwater horizon disaster observed:

> Though it is understandable that no one would wish to take ownership of the well after the blowout, the Chief Counsel's team found many instances in which nobody was taking ownership before the blowout.

References

Baker, J. et al. 2007. The Report of the BP U.S. Refineries Independent Safety Review Panel. U.S. Chemical Safety and Hazard Investigation Board, Washington.

Bartlit, F., S. Sankar and S. Grimsley. 2011. Chief Counsel's Report. National Commission on the BP Deepwater Horizon Oil Spill and Offshore Drilling. http://www.oilspillcommission.gov/sites/default/files/documents/C21462-407_CCR_for_prin_0.pdf (accessed 17 February 2011).

Borthwick, D. 2010. The Report of the Montara Commission of Inquiry. Montara Commission of Inquiry, Canberra. http://www.ret.gov.au/Department/Documents/MIR/Montara-Report.pdf (accessed 25 November 2010).

CSB (U.S. Chemical Safety and Hazard Investigation Board). 2007. Investigation Report: Refinery Explosion and Fire. Washington. http://www.csb.gov/assets/document/CSBFinalReportBP.pdf (accessed 23 November 2010).

Duncan, C. 2010. Transcript: Montara Commission of Inquiry. http://www.montarainquiry.gov.au/transcripts.html (accessed 29 September 2010).

Graham, B. et al. 2011. Deep Water: The Gulf Oil Disaster and the Future of Offshore Drilling. Report to the President. National Commission on the BP Deepwater Horizon Oil Spill and Offshore Drilling. http://www.oilspillcommission.gov/final-report (accessed 11 January 2011).

Hollnagel, E., D. Woods and N. Leverson. 2008. *Resilience Engineering Concepts and Precepts*. England: Ashgate.

chapter two

Ensuring contractor alignment with safety culture

Fiona Murfitt
DuPont Australia Pty Ltd

Contents

Introduction

An explosion killing 23 and injuring 232 workers in Pasadena, Texas (October 1989), triggered a study into the characteristic safety practices and employment conditions governing contract workers who performed maintenance, renovation, turnaround and specialised services in the U.S. petrochemical industry. The findings by the U.S. Occupational Safety and Health Administration (OSHA) identified some prevalent characteristics that are relevant across a broader base of operations utilising a contractor workforce. The study examined the following:

- The prevalence and trends in the use of contract workers
- The motivation for using contract workers
- The role of safety and health in the selection of contractors
- The safety and health training received by contract workers
- The responsibility and methods of safety oversight of contract workers, and
- The injury/illness experiences of contract and direct-hire (permanent) workers (Wells, J.C., Kochan, T.A., and Smith, M., 1991).

Some of these findings from the study include:

- The longer-term trending in the increased use of a contractor workforce
- On average, the contract workforce were younger, have less experience and education than other fully employed workers
- The leading motivator for hiring contract workers was the increased flexibility in modes of production, and therefore, there was an increased need for specialised services
- There was much variability in the extent to which plants included safety in their contractor selection criteria
- There was much variability in safety training given to contract workers vis-à-vis regular workers
- There were differences in supervisor oversight from operation to operation
- The training and experience of contract workers was less thorough, leading to increased risk of incidents in comparison to that of full time workers.

The trends and findings from this study in 1989 are still valid today. Research from some of the world's top oil and gas producers Safety Performance Indicator (OGP, 2010) has found that contractors are nearly four times more likely to be injured at work when compared to permanent employees. Of the 3,433 million work hours completed in the international oil and gas industry in 2010, 79% were attributable to contractors whilst just 21% were attributed to permanent employees. It is therefore clear that the management of contractor safety is a critical task, where the methods of planning and management in an integrated way should be better understood.

Considering these statistics, it's no wonder that there is growing need for an integrated approach to the management of contractor safety. Apart from the moral obligations, there is a strong business case for this. A strong Contractor Safety Management helps improve workers well-being, it reduces the people and financial costs of injuries and overrun, it better protects the asset, contractor effectiveness is increased, relationships are improved and it also sets and maintains acceptable safety standards throughout the organisation. The business case is made even stronger by the trend for contract organisations to differentiate based on their safety performance, as this is now being seen by some as a market differentiator providing those that manage their safety and the safety of others with a competitive advantage.

The Australian Context

Though global uncertainty has market movements have led to increased uncertainly in growth prospects, Australia is projected to continue a positive trajectory of GDP growth. This is partially buoyed by a significant investment in mining, up 13.9% from 2011 to 2012 according to the Australian Bureau of Statistics, which has insulated the country from slowdowns in the region (International Monetary Fund 2013). Much of this investment goes into large-scale capital projects, which has naturally led to high demand for skilled and non-skilled contractor labour. It is estimated that the mining sector alone will need to employ 86,000 additional workers by 2020 in order to achieve currently predicted increases in output, a 68% increase from 2008 (Molloy and Tan, 2008).

Such demand for skilled workers can often lead owner companies to create either explicit or implied pressure on contractors to just deliver the contract that is to deliver on time and to avoid delays and associated increases in cost is the most, or in some cases the only requirement. However, such pressure can also result in the unintended consequences of shortcuts being taken or work being executed at 'all cost', resulting in incidents and injuries. Not only is the injury to workers then significant, there are also impacts to production, project overrun and significant reputational damage. There is a careful balance required to deliver against goals in a planned way to meet deadlines and with driving a culture of 'just get the job done'.

Workers exerting their best efforts usually try to do the right thing. They usually do not purposely set out to be injured or hurt others. However, in Australia, the statistics suggest it will be the contractor that is more likely to be injured. According to Safe Work Australia, construction workers accounted for 11% of all serious worker claims and have a fatality rate double that of the national average (Safe Work Australia, 2012).

In addition to the obvious impact on the lives and communities of the injured worker, such incidents can result in significant financial burden for operations, project budgets and reputation. In addition to baseline compensation costs, workplace injuries and employee absenteeism can potentially slow down or halt a work schedule and if not properly mitigated, can result in a series of cascaded delays throughout the remaining timeline. Considering the already high cost of labour in Australia, companies are increasing their focus on productivity and can ill afford any further increase in cost.

Why is contractor safety important?

Owner companies hire two types of workers: their own direct-hire (permanent) employees and contractors. Permanent employees work within a defined set of systems, processes and procedures within the social environment and culture of the company. With permanent workers, a strong safety culture driven by management commitment and visible leadership, where the responsibility to care for oneself and each other is a value held by all workers, is an effective way of sustaining safety. Line managers are held accountable for the safety of employees. This should extend through to the contractor workforce.

Whilst permanent employees should work within a defined set of systems, processes and procedures within the social environment and culture of the owner company, using this exact same approach to achieve the same outcome when working with contractors can have its challenges. This is because the underlying reasons for engaging a contractor workforce and the type of contractor and the work requirements can vary considerably. This can also result in a range of cultural factors in play when considering the safety of contractors that include; 'They're only contractors', 'Contractors do all the high risk and dirty work', 'They are specialists, they know what they are doing', 'I've delegated to them and it's their responsibility to manage', 'If only the contractors were as concerned about safety as our own people are', adding to the fact much of the work may be being executed in remote locations away from supervision of any kind these are all quotes from the industry that are referenced in some form or another (both in Australia and globally) that can result in unconscious bias or mind sets that need to be addressed.

Fundamentally, the activities contractors undertake tend to be operational by nature. No matter how varied (from specialised tasks such as hot taps into live lines, erecting scaffolding or providing the necessary people for manual tasks) contractors are most often 'executing' the work. This factor, combined with the reality that contractors are often working in an unfamiliar environment, increases the likelihood of an incident occurring. The implications can be far reaching with catastrophic potential. Sadly, there are too many examples to call upon that have resulted in the loss of life and injury.

Where there is a contractor workforce, there are additional steps that need to be undertaken to integrate these workers into the culture of the organisation. This does bring with it additional challenges as there is such a variance in the reasons why the contracting workforce is engaged and with the type and length of the work being undertaken. Recognising that a contractor relationship brings with it some additional obligations is the first step in starting to manage the relationship and as such, the communications and the clarity of what is required become even more critical.

The importance of contractor safety management cannot be delegated and any approach to contractor safety management is not a stand-alone formula for success. It is important for all approaches to be integrated into the systems, structures and procedures that exist within the operations of every business.

Establishing early roles and responsibilities in the relationship between the organisation and the contractors fosters the ability to engage and clearly communicate the required objectives and standards of performance more effectively. The performance of contractors should be scrutinised not only from a financial and productivity standpoint, but also from a safety perspective. Throughout the project, there is also a role to play in setting the bar high and holding everyone to the same set of standards—both contractors and employees (from the CEO right down to operational workers).

The stronger the communication, support and encouragement, the better able the contractor will understand and meet safety performance expectations. Essentially, contractors who have a shared value for safety, and can deliver contracts safely, are also able to deliver quality work on time and on budget due to the operational discipline that good safety performance require.

Safety culture alignment

In recent decades we have seen the rise in the global energy and mining industries, which have become increasingly competitive in recent years. Public sentiment towards such large and highly visible companies now reflects the growing desire for these companies to realise their roles as corporate citizens requiring greater accountability for social responsibility, integrating safety into this their business strategy.

When safety is quoted by company leaders to be a core value, what does it really mean? In this context we define organisational culture to things like the values, beliefs and accepted behaviours that employees share through myths, stories, rituals and specialised language. Consider the idiosyncrasies of your own work community, for example, the symbolism of a corporate logo or the rituals of the Christmas party. This culture conveys a sense of identity for employees and can in turn facilitate a sense of commitment and act as a mechanism to guide and shape behaviour. Therefore, is this core value felt by the organisation at large? Is this a core value that is shared by all employees and contractors and is it exercised with true commitment and passion?

When an organisation includes safety as a part of its core value and culture, it becomes entrenched and is vitally important at both an individual and group level. The presence of a robust 'safety culture' is a good predictor of safety performance behaviours, safety knowledge, safety motivation and overall business performance.

Developing a safety culture and viewing safety as a strategic business value are the key factors in achieving business excellence today. Improved safety, including preventing injuries, saving lives and enabling a more productive workforce and more productive plants, also enhances a company's bottom line. In essence, resources are used more efficiently, employee turnover is reduced and company operations run more efficiently with enhanced profitability.

The adaptation of organisational culture to incorporate a core safety component can help equip employees with a belief in the importance of safe behaviours. This very same culture also needs to be communicated to the contractor and can also be felt by them. No longer are they adhering to safety rules because they are concerned about punishment or are anticipating reward, but because they genuinely believe it is the right way to act. A clear safety vision and policy needs to be set, and communication should be two way. There needs to be continuous safety development activities and clarification of accountability and responsibility between the owner company and its contractor.

To ensure a safe working environment, owner companies must hire contractors who can demonstrate the alignment with their own rigorous safety expectations—both before the work begins and whilst work is under way. The relationship that follows will enable owner and the contractor to interact with each other, review and audit worker safety, develop safety programs and conduct incident investigation on a joint basis.

DuPont, for example, requires the same levels of participation and adherence to site requirements for all workers, whether they are the 65,000 permanent employees or its approximately 30–35,000 contractors (hired on any given day worldwide). As such, standards, rules and expectations apply similarly to all workers. It is viewed as unethical to assign a different standard for each group. The company aims to protect the lives and quality of life of all people when they perform work at any facility, regardless of being a DuPont employee or a contractor.

This approach does not equate to a takeover of contractors' operations or safety programs by the owner company. On the contrary, contractors continue to manage and take the initiative on their own terms. But owners will measure and monitor the safety behaviour and enforce expectations as necessary. In many cases, they may also choose to participate in initiatives where learning and skills transfers are encouraged. This is encouraged by DuPont across its operations and plants in more than 90 countries.

The influence/enforcement approach is arguably the best way to manage the contractor safety risk. Based on cooperation and constant interaction, it enables both parties to focus on achieving and sustaining their commitment to keeping people, processes and their site safe.

Recognising that there are additional steps and obligations when working with a contractor workforce, DuPont uses an integrated six-step approach to deliver and sustain the management of contractors that focuses on selecting, engaging, communicating, reviewing, training and managing. The use of these six steps has been proven effective many times for DuPont's own engagements with contractors and also for clients of Dupont's Consulting Business, DuPont Sustainable Solutions, who have deployed this process. However, any approach to contractor safety management is not a stand-alone formula for success. It is important for all approaches to be integrated into the systems, structures and procedures that exist within the operations of every business.

Contractor safety management: the Integrated approach

In order to deliver improved safety outcomes for contractors, DuPont developed an integrated approach to contractor management that has proven successful in decreasing the frequency of injuries and accidents amongst its contractors and with assisting the delivery of projects to be on time and on budget. The system is composed of six complementary processes:

(1) the selection of contractors with satisfactory safety records, (2) the inclusion of safety standards in contractual obligations, (3) clarification of expectations upon award of bid, (4) orientation and training of contractor teams, (5) monitoring of safety activities, and (6) a post-contract evaluation to assess success and lessons learned.

The first four steps are considered front-end loading, and should be the focus for owner companies, as it is at this stage, the owner can best define the relationship with the contractor, and thus add the most value.

1. ***Contractor Selection*** As previously discussed, the contractor is ultimately responsible for ensuring the safety of its employees. By selecting a contractor with an exemplary safety record, it is much more likely that the work will be performed safely. There is also a collection with this and with projects finishing on time and on budget.

   An effective contractor selection process involves evaluating the contractors on their past safety performance. The contracting organisation must seek metrics or leading and logging safety data like the LTI rate (LTIFR) or Total Recordable Frequency Rate (TRR) and recordable rates or similar statistics from the regulatory body. However, in order to supplement past performance records or these lagging indicators, the owner company must also do a safety competency assessment of the contractor. It is also important that the safety department/professional along with operations of the owner company are involved in the decision-making process to ensure that safety is given equal weight with other factors, such as cost discipline. Other hurdle criteria could be the level to which operating discipline is enforced, and also whether the company employs a strong culture of visible interactions between management and the operations team, along with an audit regime.

   Some common mistakes during this step are placing more weight on past records or written progress rather than on competency and safety management; selecting the low bid without examining what the contractor will actually deliver for that price; and by utilising internal decision makers who do not have right skill sets to make informed choices regarding safety and operations dimensions of the contractor selection process.

2. ***Contract preparation*** The next, and most important, step in the contractor safety management process is to prepare the contract. The contract establishes the rules and conditions in which the contractor will operate. It is at this stage that the owner company is able to create a structure to ensure that safety is fully integrated into operations, thus making it the main point of leverage when interacting with the contracting party during execution.

   When preparing the contract, all contract terms and conditions must clearly document safety parameters, such as expectation for performance, behaviour, standards and capabilities of key personnel. The parameters must be targeted to the scope of work and be clearly related to a hazard analysis performed by the owner. Furthermore, the contract must specifically place responsibility and accountability for contractor and sub-contractor safety with the main contracted party. On a more practical level, the contract must also define the communication channels through which the contractor will disseminate knowledge pertaining to safety, and also stipulate that sufficient resources be made available for orientation and training, including specific regulatory training requirements.

   The owner company must involve all constituents in the development of contract language, including safety resources and field contract administrators, and must clearly identify roles and responsibilities for developing contract packages. A carefully developed contract not only helps set performance expectations but also serves as a road map to guide the relationship. Finally, it must cover planning and documentation requirements.

3. ***Contract award/establish expectations and standards*** On awarding the contract, the owner company must communicate and test understanding of safety expectations that are defined in the contract. The owner must not assume that contractors will read and understand all safety requirements and must walk supervisors through the rules. This discussion should include customised commentary on behalf of the owner company that provides detailed information on the specific scope of on-site hazards and expected hazard controls. At this point roles and responsibilities must be communicated, including responsibility to conduct meetings, responsibility to

discuss contract safety requirements, ensuring that those who have the role of reviewing safety requirements have the right knowledge and training to adequately present and answer any questions. In essence, the contracting party must be fully aware of their role in ensuring safety.

4. ***Orientation and training*** Whilst the contracted party maintains primary responsibility and accountability for contractor safety, the owner company also has a role in ensuring contractor safety. The owner should utilise their own knowledgeable, experienced employees to provide effective orientation and safety training. Attributes of an effective orientation include a qualified instructor/presenter, proper explanation of the hazards and specific work environment, as well as a system that measures understanding of safety requirements.

   Common mistakes made by owner companies during this step are the failure to customise the orientation to suit audience, the treatment of orientation as trivial, and rushing through it, a disconnected orientation not driven by results, and operating under the perception that orientations are single events rather than an ongoing effort. An effective orientation program is a foundation for the desired safety performance at the contractor level.

5. ***Monitoring safety activities*** In order to ensure compliance to safety rules, the owner must develop a robust system for monitoring safety activities, and have already defined this program within the context of the contract. Again, the main responsibility for monitoring lies with the contracting party, yet a robust enforcement system is necessary to complement the activities of the contracting party. Effective monitoring calls for a partnership with the contractor rather than an adversarial relationship. Key elements of such a system would be formal safety audits and inspections, incident investigations, continuous updating of job plans and periodic review of safety systems. The owner and contractor must jointly develop periodic safety meeting materials, conduct periodic status review meetings and targeted pre-job safety plan reviews and must investigate any incident or accident for potential lessons. A common pitfall is emphasis and effort focused at this step to achieve results, rather than front-end loading process—the first four steps.

6. ***Evaluate safety performance against contractual expectations*** In the larger context of contractor safety management, this step serves as a tool for continuously improving the process. In this step, the owner must critique contractor performance against contractual expectations, and also provide detailed, constructive feedback to

the contractor to facilitate improvement. Where expectations were not met, records must be updated to reflect this. Contractors not performing sufficiently relating to safety should not be selected for further contracts. Some of the review areas are: injuries/incidents, workers compensation and general liability claims, lost workday cases, TRFR, property and vehicle damage.

Implications for field application

The application of this integrated contractor safety management process has led to significant improvements in contractor safety performance and the delivery of on-schedule and to-budget projects, dispelling a myth that safety costs money. This is both true for DuPont and its clients.

Within DuPont, the application of this process can be best exampled from the Cooper River expansion project in the United States. The recent expansion of the Cooper River facility, located north of Charleston, South Carolina, was the centrepiece of a multi-phase, multi-year Kevlar® production expansion that initially increased global Kevlar production capacity by more than 25%. This expansion was designed to help meet the growing global demand for Kevlar®,* a para-aramid brand fiber for industrial and military uses. This expansion (along with another smaller expansion at the Spruance Plant), represented the largest single investment in Kevlar and the largest capacity increase since the fiber was introduced in 1965.

During the construction of this Cooper River site near Charleston, South Carolina, a rigorous contractor safety management system delivered 3.1 million man-hours of construction work without a lost-time injury, despite having 400–800 contractors on site at all times throughout the project cycle. The construction also delivered an operational plant that was delivered on schedule and in accordance with the allocated budget.

A client, Titan Cement, one of the leading cement producers in the world, decided to build a new cement manufacturing line in Beni Suef, Egypt. The company was investing 150 million euros, but had no local staff to supervise the project on-site. They therefore had to rely on contractors, who sub-contracted the work to local Egyptian firms. By implementing a robust contractor management system, and with the help of DuPont, the company managed to run the project without a lost-time injury in the 6.5 million man-hours worked. Further, the project was delivered on time.

In Australia, DuPont has worked with an engineering services company in Queensland which is a contracting company to the resource

* Kevlar® is registered trademark of E. I. duPont de Nemours and Company.

sector that has proactively sought to differentiate itself with their competitive advantage, its record in safety. Between 2004 and 2010, the company reduced its recordable injury rate by 85%, with the help of DuPont.

DuPont has also worked with Australian construction giant, which was nationally recognised for its unique approach in including subcontractors as part of its overall safety management system. They achieved a reduction of 81% in its total injury frequency rate from 2008 to 2011, whilst one of its subcontracting companies achieved a 73% reduction in just 12 months. 'Our safest projects are our best performing projects', says their construction manager, and indeed it is no coincidence that the safest projects are also those that are completed to specification, on time and within budget.

In these cases, the improved safety record has allowed the contracting companies to achieve significant improvements in overall business performance and to reinforce their reputation for safety as a core value.

Conclusion

There are several cost associated with safety failures—direct costs like workers' compensation, medical/hospital costs, property losses and liability losses, and indirect costs like administrative/investigative time, negative publicity, cleanup, disrupted schedules, loss of productivity, owner civil liability, legal fees/awards/settlements, overtime costs, replacement/repair of equipment and training of replacement employees. On the other hand, there are several benefits associated with improvement in safety performance: fewer workers are injured; cost associated with contractor incidents and schedule delays is reduced; and productivity losses are avoided, thus protecting owner assets, improves owner/contractor relationships, increasing contractor effectiveness and reducing negative publicity associated with injuries.

Whilst companies and contractors have the same safety objective—to prevent all injuries and incidents—it is ultimately the company that determines the level of safety delivered in the workplace. Companies that have implemented safety management systems that help create and sustain safety culture have seen dramatic reductions in injuries and incidents amongst contractors. An output of this is also an increase in productivity and an improvement in employee morale. As discussed in the body of this chapter, to deliver improved safety outcomes for contractors, DuPont developed an integrated approach to contractor safety management that has proven successful in decreasing the frequency of injuries and incidents amongst its contractors. The system is composed of six complementary processes: (1) the selection of contractors with satisfactory safety records, (2) the inclusion of safety standards in contractual obligations, (3) clarification of expectations upon award of bid, (4) orientation and training of contractor teams, (5) monitoring of safety activities, and (6) a post-contract evaluation to assess success and lessons learned.

Any approach to contractor safety management is not a stand-alone formula for success. It is important for all approaches to be integrated into the systems, structures and procedures that exist within the operations of every business. To ensure success, companies must first develop an appropriate context for contractor safety by familiarising themselves with safety issues that affect contractors specifically, demonstrating a visible management commitment and developing a strong safety culture based on individuals valuing safety and operational discipline and where there is a collective responsibility for safety.

In terms of leadership, it is vital that leaders, managers and supervisors demonstrate a strong commitment to safety. Employees must feel as though this commitment is genuine and deep and is what DuPont refers to as 'felt leadership'. Safety must be considered and integrated into significant management decisions, just as quality, cost and productivity are considered. To the second point, structure, it is important that key members of the organisation's safety program be deployed strategically throughout the company. This furthers the ability of employees to be directly involved and truly engaged in delivering safe outcomes. Moreover, line managers must be held accountable for safety performance. The third requirement of a strong safety culture is processes and actions, which dictates that safety be streamlined into each of the company activities. An important element of these processes relates to communication channel. To ensure consistent, clear communication on safety matters, platforms for communication must be developed and maintained.

A safety culture can be achieved by ensuring organisational commitment, management involvement, employee empowerment and appropriate systems for reward and reporting. These mechanisms can help influence the thoughts and beliefs of employees through contextual and social influence and there is a critical role in ensuring that contractor management and their safety is integrated into this.

Ultimately, 'You get the level of safety that you demonstrate you want' (DuPont proverb).

References

International Monetary Fund (2013). *World Economic Outlook 2013*. Washington DC.

Molloy, S. and Dr. Yan, T. (2008). *The Labour Force Outlook in the Australian Minerals Sector; 2008 to 2020*. National Institute of Labour Studies, Flinders University, Adelaide, Australia.

OGP Safety Performance Indicator. 2010, Report nos 455 may 2011 (updated June 2011). Retrieved from http://www.ogp.org.uk/pubs/455.pdf,

Safe Work Australia (2012). *Construction Fact Sheet*. Canberra: Safe Work Australia.
United Nations Department of Economic and Social Affairs (2009). *Trends in International Migrant Stock: The 2008 Edition*. New York: United Nations Department of Economic and Social Affairs.
Wells, J. C., Kochan, T. A. and Smith, M. (1991). Managing Workplace Safety and Health: The Case of Contract Labor in the U.S. Petrochemical Industry. Final Report to the Occupational Safety and Health Administration. Washington, D.C: U.S. Department of Labor.

chapter three

Bridging the safety gap: Engaging specialist contractors and the duty to consult, coordinate and cooperate

Olga Klimczak
LLB (Hons.) BA (Hons.)
LLM Candidate, University of Melbourne

Contents

Introduction

Contemporary work environments and work arrangements are often complex. The performance of work often involves multiple parties working across the same or various sites, who have different levels of expertise, knowledge and control over the performance of work by workers. Additionally, parties often engage specialist contractors with expertise in relation to particular work activities.

These complexities can give rise to gaps in relation to safety-related matters—gaps in communications between different parties working in close proximity at the same site, where the performance of one party's work may impact on the health and safety of other workers, or gaps that each party assumes that another has dealt with certain safety-related issues, and in fact no one does. Further, in the context of parties engaging specialist contractors, there can be confusion regarding the extent to which these parties can simply rely on the specialist contractor, or whether these parties are required to take additional steps, such as supervising, or issuing safety-related directions to, the specialist contractor in order to meet their obligations under workplace health and safety laws. This is compounded by the fact that there are multiple legal regimes that apply in this context, with multiple, concurrent duty holders owing overlapping duties, and one duty holder potentially owing a number of different duties.

Some recent Australian cases have considered the key legal principles that apply in respect of workplace health and safety in situations where specialist contractors have been engaged. Moreover, as part of the national review into model Occupational Health and Safety Laws, the advisory panel considered some of the potential gaps identified above and proposed a new duty on duty holders to consult, co-operate and co-ordinate with other duty holders (Stewart-Crompton, Mayman and Sherriff 2008, 2009). This duty has been adopted in the model workplace health and safety legislation and has been implemented in some jurisdictions (Dunn and Chennell 2012).

In this chapter, I analyse the recent cases regarding safety involving parties who have engaged specialist contractors, and explore the limitations on parties relying on expert contractors to meet their own safety-related duties. I argue that these recent cases have re-affirmed that, even in this context, parties must take a proactive approach to workplace health and safety issues, and may not be able to rely on specialist contractors in relation to safety-related matters. Additionally, the new duty to consult, co-operate and co-ordinate activities with other duty holders is significant in that it expressly recognises the importance of parties engaging in a dialogue with other duty holders in relation to safety matters, so as to minimise the potential of the gaps identified above occurring.

I will begin with an overview of the legal framework regulating workplace health and safety, particularly focusing on the context where parties engage specialist contractors. I will compare the Victorian Occupational Health and Safety Act 2004 (pre-model legislation) with the new regime under the Workplace Health and Safety Act 2011 (Cth) (model legislation).

In the second part, I review a number of Australian cases involving prosecutions of parties who had engaged specialist contractors for an alleged failure to meet their safety-related duties. My focus will be on the question of how courts have approached the issue of breach and reasonable practicability in this context. I will contend that the key factors in the courts' assessment of whether further steps are reasonably practicable under the circumstances, other than simply relying on the expert contractor, are, first, the respective control, expertise and knowledge of the parties in relation to the hazard and measures to eliminate or reduce it and, second, whether the party otherwise has some notice or knowledge of a problem regarding the expert contractor's work or potential hazards (where the matter falls outside its expertise). I conclude that a clear, recurrent theme is that the gaps previously referred to often arise in the context of engaging expert contractors, with tragic consequences.

In the final part of this chapter, I will explore whether the new duty to consult, co-ordinate and co-operate in the model legislation is likely to assist in addressing these gaps, and whether it will have a positive impact in the context of managing contractor safety issues. As part of this analysis I will consider the background to the proposed new duty and its content. I conclude that although this new duty has not had much attention, its importance in promoting a safety dialogue between duty holders should not be underestimated.

Legislative context

Introduction

There are a number of different laws which regulate health, safety and welfare in the work context. These include the 'common law of contract' (implied term that an employer has a duty to provide a safe workplace) (Creighton and Stewart, 2010) and negligence (duty of care of employers, principals, occupiers and others, and the vicarious liability of employers for the negligent acts of employees resulting in harm to third parties; Sappideen and Vines, 2011), occupational health and safety statutes imposing criminal liability (both general and industry-specific legislation, e.g., mining, rail, aviation, oil and gas, maritime; Dunn and Chennel, 2012), and statutes dealing with compensation and rehabilitation of injured workers (Johnstone, Bluff and Clayton, 2012).

In relation to the statutory duties under the pre-model and model legislation, the key issues are:

- The identity of the duty holders;
- The scope of the duty (in particular, to whom the duty is owed); and
- The nature of the duty (that is, what is required).

Duty holders and scope of their duties

Parties who engage specialist contractors may owe duties in multiple capacities to various persons under the pre-model and model legislation, as set out below.

Duty of employer to employees

First, if the party is an employer, it will owe a duty of care to its employees under both the pre-model and model legislation. The traditional paradigm of work involving an employer served by long-term, permanent, full-time employees (known as the 'Harvester' model) has underpinned workplace health and safety regulation for much of the 20th century and the first decade of the 21st century (Dunn and Chennel, 2012; Johnstone, Bluff and Clayton, 2012).

The pre-model legislation imposes safety duties on employers to, so far as is reasonably practicable, provide and maintain for their employees a working environment that is safe and without risks to health (Occupational Health and Safety Act 2004 (Vic) s 21[1]). Further guidance on what this means is provided by way of a non-exhaustive list of failures that would amount to a contravention of the duty (Occupational Health and Safety Act 2004 (Vic) s 21[2]). This list is broadly consistent with the common law duty of care. Employers must also, so far as is reasonably practicable, monitor the health of employees and the conditions at any workplace under its management and control, and provide safety-related information to employees (Occupational Health and Safety Act 2004 (Vic) s 22).

In contrast, the primary duty holder under the model legislation is the 'person conducting a business or undertaking' (PCBU; Workplace Health and Safety Act 2011 (Cth) s 19[1]), consistent with the recommendations of the review panel (Stewart-Crompton, Mayman and Sherriff, 2008). The review panel considered changes to workplace arrangements and relationships in recent times, and noted that the traditional category of duty holder, the employer, was no longer the dominant work relationship in connection with contemporary work. A number of commentators have also recognised that there have been significant changes in relation to the way work is organised, including that work relationships and work arrangements have been impacted by globalisation and resulting competitive pressures (see, for example, Johnstone and Tooma, 2012; Johnstone,

Bluff and Clayton, 2012; Bluff, Gunningham and Johnstone, 2004; Reeve and McCallum, 2011; Scott, 2005). Outsourcing, contracting, labour hire, complex supply chains and other forms of precarious work are now increasingly common. Further, work is no longer fixed to specific locations or premises, with the development of technologies that allow workers to perform work from home, during travel or commuting, in remote locations, or otherwise away from the employer (Tooma, 2016). These changes require different approaches to health and safety regulation to achieve the objective of ensuring safe working environments, so far as is reasonably practicable. This was the rationale for adopting the PCBU as the primary duty holder under the model legislation. The term *PCBU* is defined in the model legislation and is to be interpreted broadly (Safe Work Australia, 2011b). However, it is clear that a PCBU would include the traditional employer duty holder.

Duty of principal to contractors

Second, a party engaging a specialist contractor will owe duties to the specialist contractor or its employees. Further, if other contractors are also engaged, the party will also owe duties to those contractors and their employees. In this context, there is the additional element of coordinating the various contractors' activities to ensure that the potential risks from having multiple contractors performing various activities are managed.

The pre-model legislation extends the employer's duty to workers who are deemed to, in effect, be treated as employees for the purposes of the employer's duty. This is achieved by providing that references to 'an employee' include references to an 'independent contractor' engaged by an employer and any employees of the independent contractor, in relation to matters over which the employer has control (or would have control but for an agreement purporting to remove or limit the control) (Occupational Health and Safety Act 2004 (Vic), s 21[3]). The terms 'independent contractor' and 'engaged' are not defined, but have been interpreted widely by courts to include subcontractors, and further subcontractors in contracting arrangements involving multiple layers of contractors (*R v ACR Roofing Pty Ltd*, 2004, pp. 187, 188–189, 202–208).

The pre-model legislation limits the duty to non-employee workers by utilising the concept of 'control'. The effect of this is that the party will not owe a statutory duty to the specialist contractor's employees in relation to matters over which it does not have control. This is similar to the duties under the common law, which also refer to control, supervision and vulnerability in determining whether a non-delegable duty should be imposed on a principal (*Stevens v Brodribb Sawmilling Co Pty Ltd*, 1986; *Crimmins v Stevedoring Industry Finance Committee*, 1999; *Leighton Contractors Pty Ltd v Fox*, 2009; *Pacific Steel Construction Pty Ltd v Barahona*, 2009).

However, Johnstone and Tooma (2012) have noted that workers who are not employees, or deemed employees, would nevertheless fall within the scope of the employer's duty to other persons (discussed further below).

The model legislation takes a different approach in that the scope of the duty is not limited by the concept of control (although control is relevant to the issue of reasonable practicability discussed further below). The review panel were concerned that limiting the duty by reference to 'control' may focus the consideration on whether there is a duty at all, rather than determining the extent of the duty, that is whether the duty holder had done all that was reasonably practicable (Stewart-Crompton, Mayman and Sherriff, 2008). As will be seen in the discussion in the next part, one of the arguments in cases involving principals who have engaged specialist contractors is that the principal has no duty in relation to the particular matter, because of a lack of control over the specialist work (for example, *Reilly v Devcon Australia Pty Ltd*, 2007). The review panel considered that by legislating a broad duty to all workers regardless of how they are engaged, the focus would not be on whether a duty is held, but rather, what steps the duty holder must take to discharge its duty (Stewart-Crompton, Mayman and Sherriff, 2008).

Section 19(1) of the model legislation provides that a PCBU's duty to workers relates to workers who are either engaged or caused to be engaged, or whose activities in carrying out work are influenced or directed, by the duty holder, whilst the workers are at work in the business or undertaking of the duty holder. The wording of the section makes it clear that a direct contractual relationship is not necessary. Further, the model legislation goes beyond the concept of persons engaged or caused to be engaged by the duty holder, to include circumstances where even if the worker was not engaged by the duty holder, the duty holder nevertheless has capacity to direct or influence the worker's work activities (Johnstone, Bluff and Clayton, 2012). These broad provisions would most certainly apply a duty to a PCBU engaging a specialist contractor, or indeed any independent contractor to perform work that could readily be done by its own employees. In this regard, the term *worker* is broadly defined in Section 7 of the model legislation as a person who carries out work in any capacity for a PCBU. The definition sets out a non-exhaustive list of examples of workers, including those covered by the pre-model legislation, namely, employees, contractors and subcontractors, and labour hire staff. It also expressly includes outworkers, apprentices or trainees, work experience students, and volunteers (Workplace Health and Safety Act 2011 [Cth], s 7).

Duty to other persons

Third, the party engaging the specialist contractor may also owe duties to other persons, such as the general public, visitors to premises, or where there is no relationship between the party and the person (e.g., other workers in close proximity).

Under the pre-model legislation, an employer or self-employed person owes a duty to other persons to ensure, so far as is reasonably practicable, that those persons 'are not exposed to risks to their health or safety arising from* the conduct of the undertaking of' the duty holder (Occupational Health and Safety Act 2004 (Vic), ss 23, 24). This duty is cast slightly differently in other pre-model legislation (see, for example, Occupational Safety and Health Act 1984 (WA), s 21). Under the model legislation, the wording is similar, although slightly different, with the potentially narrower 'put at risk' being used instead of 'exposed' (Tooma, 2012; Johnstone, Bluff and Clayton, 2012).

Other duties

Finally, duties are also placed on self-employed persons (Occupational Health and Safety Act 2004 (Vic), s 24; Workplace Health and Safety Act 2011 (Cth), s 19(1)), those parties having control or management of workplaces, and certain 'upstream duty holders' (designers, manufacturers, suppliers, importers, installers and erecters, etc.) (Occupational Health and Safety Act 2004 (Vic), ss 26–31; Workplace Health and Safety Act 2011 (Cth), ss 20–26).

Nature of general duty

Concurrent duties

I have already noted above that a party engaging a specialist contractor may have multiple, simultaneous duties to various persons. This is expressly recognised in the model legislation, which provides that a person can have more than one duty by virtue of being more than one class of duty holder (Workplace Health and Safety Act 2011 (Cth), s 15). This forms part of the model legislation's express statement of interpretative principles to apply in approaching the duties (Workplace Health and Safety Act 2011 (Cth), ss 13–17).

These express interpretative principles also acknowledge that more than one person can concurrently have the same duty in relation to the same matter (Workplace Health and Safety Act 2011 (Cth), s 16(1)), and that, in these circumstances, each duty holder retains responsibility for their duty and must discharge their duty to the extent to which they have the capacity to influence and control the matter (or would have but for an agreement purporting to limit or remove that capacity) (Workplace Health and Safety Act 2011 [Cth], s 16[2]). Further, whilst this is not expressly stated in the pre-model legislation, so much is clear from multiple parties being prosecuted in relation to the same circumstances (see, for example,

* The Workplace Health and Safety Act 2011 (Cth) uses from 'work carried out as part of'.

Tobiassen v Reilly, 2009; *Reilly v Devcon Australia Pty Ltd*, 2008; *Kirwin v Laing O'Rourke (BMC) Pty Ltd*, 2010; *Kirwin v Pilbara Infrastructure Pty Ltd*, 2012). Therefore, in the context of parties engaging specialist contractors, other persons will likely have concurrent duties, including the specialist contractor, other employers or contractors, and workers. Each of these other persons, as well as the party engaging the specialist contractor, must ensure that they discharge their relevant duties to the requisite degree.

In this context, it is important to note that employers must provision for the fact that employees may act inadvertently or without reasonable care, and this will not be a mitigating factor. As was said in *R v Australian Char Pty Ltd* (1999, pp. 848–849):

> But long experience has shown that employees do sometimes act inadvertently or without due care for their own safety. It is in that context that an employer must guard against such acts or omissions as may foreseeably cause injury.

However, as will be discussed in the next part, the West Australian Supreme Court rejected WorkSafe WA's argument that a similar approach should be taken in the context of specialist contractors, on the basis that this was not reasonably foreseeable and it was not reasonably practicable to do anything more than rely on the specialist contractor.

Personal and non-delegable duties

Perhaps the most important feature of the safety-related duties is that they are personal and non-delegable. The concept of non-delegable duties in the common law context was explained by Mason J in *Kondis v State Transport Authority* (1984, p. 672) as follows:

> … the duty is of such a nature that its performance cannot be delegated to a contractor on the footing that delegation to a competent contractor is a sufficient compliance with the duty … the employer... must ensure that reasonable care and skill is exercised in relevant respects.

The statutory duties under the pre-model legislation are generally non-delegable. The same position applies under the model legislation, except that rather than leave it to case law, Section 14 expressly provides that 'A duty cannot be transferred to another person'.

However, as will be discussed in the next part, because the safety duties under the pre-model and model legislation are limited by the qualifier of reasonable practicability (discussed next), it may be said that the outcome in some of these cases is that the party engaging the specialist contractor has, in effect, delegated its duty in relation to safety matters falling within the province of the specialist contractor's expertise, and in respect of which the party has no expertise.

Qualification of reasonable practicability

Finally, the last issue to discuss in this context is that under both statutory regimes, the duty is not to ensure that accidents never happen, and is limited by what is reasonably practicable in the circumstances (*Holmes v R E Spence & Co Pty Ltd*, 1993). This is important in the context of both the pre-model legislation and the model legislation imposing criminal liability on duty holders who fail to comply (Occupational Health and Safety Act 2000 [Vic], ss 21–24; Workplace Health and Safety Act 2011 (Cth), Pt. 2 Div. 5). Prosecutions under these statutes may only be brought by the relevant regulator, who must establish the contravention beyond reasonable doubt.

Whilst cases have warned against looking at issues with the benefit of hindsight or the 'wisdom of Solomon', they nevertheless require parties to take 'an active, imaginative and flexible approach to potential dangers in the knowledge that human frailty is an ever-present reality' (*Holmes v R E Spence & Co Pty Ltd*, 1993). This approach applies equally in the context of parties engaging specialist contractors.

In this context, both statutory regimes recognise a hierarchy of controls, being that the risk should be eliminated so far as is reasonably practicable; and if it is not reasonably practicable to eliminate, the risk, only then should measures be put in place to minimise the risk, so far as is reasonably practicable (Occupational Health and Safety Act 2004 (Vic), s 20[1]; Workplace Health and Safety Act 2011 [Cth], s 17). Further, duty holders must be vigilant, and it is not necessary for there to be an injury or incident for there to be a breach of the safety duties. Rather, putting a worker or person at risk where there was a reasonably practicable measure to eliminate or reduce the risk will, of itself, be sufficient to constitute a breach of the statutory duty (*R v ACR Roofing Pty Ltd*, 2004; *R v Australian Char Pty Ltd*, 1999).

Turning then to what is meant by the phrase 'reasonably practicable', both the pre-model and model legislation define the phrase, but in different ways. Under the pre-model legislation, Section 20(2) provides that regard must be had to the following matters in determining what is or was reasonably practicable in relation to ensuring health and safety:

- the likelihood of the hazard or risk concerned eventuating*;
- the degree of harm that would† result if the hazard or risk eventuated;

* The Workplace Health and Safety Act 2011 (Cth), s 18 instead uses the word 'occurring'.

† The Workplace Health and Safety Act 2011 (Cth), s 18 instead uses the word 'might'.

- what the person concerned knows, or ought reasonably to know, about the hazard or risk and any ways of eliminating or reducing the hazard or risk. This has been referred to as the 'state of the knowledge' (see, for example, Occupational Safety and Health Act 1984 (WA), s 3);
- the availability and suitability of ways to eliminate or reduce the hazard or risk; and
- the cost of eliminating or reducing the hazard or risk.

The model legislation is in very similar terms, but the above matters are prefaced with a statement that reasonably practicable 'means that which is, or was at a particular time, reasonably able to be done in relation to ensuring health and safety, taking into account and weighing up all relevant matters including' (Workplace Health and Safety Act 2011 [Cth], s 18) the above factors. Further, the key difference is in relation to the last factor, cost. Under the model legislation, the cost of control measures must be 'quarantined' in the assessment, and left until last, 'after assessing the extent of the risk and the available ways of eliminating or minimising the risk' (Workplace Health and Safety Act 2011 [Cth], s 18[e]). As part of this analysis, the party must determine whether the cost is 'grossly disproportionate to the risk' (Dunn and Chennel, 2012; Johnstone and Tooma, 2012).

In weighing up the above factors and determining what measures are reasonably practicable in any given situation to eliminate or reduce hazards, parties are effectively called on to exercise a value judgment. As Gaudron J recognised in *Slivak v Lurgi (Australia) Pty Ltd* (2001) 'the question whether a measure is or is not reasonably practicable is one which required no more than the making of a value judgment in the light of all the facts' (*Slivak v Lurgi (Australia) Pty Ltd,* 2001, pp. 322–323). Additionally, as Her Honour noted in that case, and as was also re-affirmed recently by the High Court (*Baiada Poultry Pty Ltd v The Queen,* 2012), the phrase 'reasonably practicable' is narrower than 'physically possible' or 'feasible' (*Slivak v Lurgi (Australia) Pty Ltd,* 2001, pp. 322–323; *Edwards v National Coal* Board, 1949, p. 747).

Whilst not inconsistent with the assessment of what is reasonably practicable, an alternative approach is one of risk management (Johnstone, 1999; Johnstone and Bluff, 2005). However, the legislature has not taken the opportunity in the model legislation to prescribe such an approach, something which has been criticised by Johnstone and Tooma (2012).

In the context of parties engaging specialist contractors, one of the critical questions is what is it reasonably practicable for the party to do in respect of risks and hazards associated with the performance of work by the specialist contractor? The approach of the courts to this issue is discussed in the next part.

Cases involving expert contractors

Introduction

Parties who engage specialist contractors often struggle with the extent to which the safety duty requires them to take any additional steps or whether they can rely on the specialist contractor in relation to safety matters arising out the specialised work activity. This part discusses a number of Australian cases that have considered the issue of what is required of a party who has engaged a specialist contractor in order to discharge its statutory safety duty. In all of the cases discussed below, it is clear that there was a gap in knowledge in relation to the relevant hazard—the specialist contractor either made an error or omission, or failed to ensure a safe work method, and the party engaging the specialist contractor also did not take any other positive steps, such as issuing safety directions or supervising the specialist contractor, or engaging further experts to do so. Assumptions were made; reliance was placed on the specialist contractor. However, as will be seen, in some of these cases, the only reasonably practicable thing for the party to do was to rely on the specialist contractor. In others, the parties breached their duties by failing to take additional measures.

The question of what is reasonably practicable is a matter of fact and degree in each case (*R v Associated Octel Ltd*, 1994, p. 1062). However, the cases discussed below suggest that parties who engage specialist contractors must be vigilant and take a proactive, creative and flexible approach. A useful starting point in determining what is expected of duty holders is Steytler J's judgment in *Hamersley Iron Pty Ltd v Robertson* (1998). The following principles can be extracted from the judgment. First, if a particular task is necessary in order to discharge a duty holder's safety duties, and it has no expertise in that area, it must call on the expertise needed, in order to discharge its duties (if that is reasonably practicable; *Hamersley Iron Pty Ltd v Robertson*, 1998, p. 20). This is consistent with taking a proactive approach to safety matters. Second, a duty holder must assess whether it reasonably appears that the contractor it engages is competent at the task and is performing it carefully and safely (*Hamersley Iron Pty Ltd v Robertson*, 1998, p. 20). If not, it may be required to engage additional experts or to give directions to address any risks, if reasonably practicable, to discharge its duty. Third, the objective relative expertise of the parties, and what a reasonable person in the duty holder's position is expected to know, is relevant in determining what is reasonably practicable, as this goes to the 'state of the knowledge' aspect of the definition of reasonably practicable (*Hamersley Iron Pty Ltd v Robertson*, 1998, p. 20).

In relation to this last point, the courts' approach generally differs depending on whether the hazard or control measures are ones falling within or outside the knowledge, control or expertise of the duty holder.

Matters within duty holder's control, knowledge or expertise

The High Court has observed that, in many cases, questions of safety and practicability raise issues of common sense rather than special knowledge (*Chugg v Pacific Dunlop Ltd,* 1990, p. 260). If the hazard and control measures are within the scope of the duty holder's expertise, control and knowledge, then it would ordinarily be reasonably practicably for it to exercise that control and supervise the specialist contractor in relation to that matter or issue relevant safety directions. In this regard, duty holders cannot leave the expert to do their work and rely on the specialist contractor to discharge the duty holder's duty. This is notably different to the position under common law as it relates to a principal's duty to specialist contractors (cf. *Leighton Contractors Pty Ltd v Fox,* 2009).

In *R v Associated Octel Ltd* (1994), Stuart-Smith LJ recognised that ' … there are cases where it is reasonably practicable for the employer to give instructions how the work is to be done and what safety measures are to be taken' (*R v Associated Octel Ltd,* 1994, p. 1062). In this context, Australian courts have distinguished between matters within the control of the party engaging the specialist contractor and those outside its control. An example is the case of *R v ACR Roofing Pty Ltd* (2004). In that case, ACR Roofing Pty Ltd (ACR) was engaged to install the roof of an extension to an existing commercial building in Port Melbourne. ACR engaged an expert crane company to lift packs of new steel roof sheets onto the roof purlins, so that ACR could then install the sheets. As the lift was under way, the crane touched overhead powerlines whilst the dogman was handling a pack of roof sheets suspended from the crane, causing him to be electrocuted and fall dead from the roof purlins to the concrete floor below. The death was caused by the electrocution. However, no safety mesh was erected below where the dogman had been working.

ACR was charged with failing to ensure adequate fall protection and failing to eliminate or alleviate the risk of electrocution. It argued that it was not in control of the lift, as it had engaged experts (the crane company) to carry out the work in which ACR had no expertise, and because it was industry practice for the crane crew to take control of the site once they arrived, with the dogman invariably being in control of the lift (*R v ACR Roofing Pty Ltd,* 2004, p. 212).

However, Nettle JA distinguished the siting of the crane and method of lifting, being matters within the expertise of the crane company and dogman, and the erection of the safety mesh, which ACR would have installed for its own employees (*R v ACR Roofing Pty Ltd,* 2004, pp. 212–213). The fact that the crane company was an expert contractor and the deceased was an expert dogman was relevant in the sense that it could hardly be said that ACR had control over, or that it was reasonably practicable for ACR to do much, if anything, about the matters going to the risk of electrocution (*R v ACR Roofing Pty Ltd*, 2004, p. 213). However, Nettle JA stated that 'the erection of the safety mesh was a matter of a different order. The need for it was well known to ACR and in any event was obvious' (*R v ACR Roofing Pty Ltd*, 2004, p. 213).

Similarly, in *Candetti Constructions Pty Ltd v Fonteyn*, (2012) Candetti Constructions Pty Ltd (Candetti) was convicted for a failure to ensure a safe system of work in relation to the fall of an expert contractor through an unguarded opening. Candetti was the project manager of construction work at a Woolworths distribution centre. It had engaged Ace Systems Pty Ltd (Ace) to install cool room panels in the premises. In order to install the roof panels, a hole was made in the ceiling and scissor lift used to transfer panels through the hole and into place. The hole was ordinarily protected by orange mesh bunting around its edges, but this needed to be removed when transferring the panels into the ceiling space. Only Ace knew when this was to occur and only Ace had control over the scissor lift. At the time of the accident, the bunting had been removed. Following the accident, Candetti re-installed edge protection around the opening and directed Ace that it was not allowed to remove it. Additionally, Candetti placed a person in the ceiling permanently to monitor the opening.

The magistrate found that whilst Candetti did not have control over the actual installation of the cool room panels, the specialist work that Ace, the specialist contractor, had been engaged to perform, Candetti did have control over general safety at the site, and this included the provision and maintenance of secure fences, covers and other forms of safeguarding around openings in the ceiling space through which workers could fall (*Candetti Constructions Pty Ltd v Fonteyn*, 2012, p. 251). On appeal, Candetti argued that it did not have control over the opening, and rather, at Ace's request, it had left control over the opening to Ace to allow it to move the panels into the ceiling space for a purpose specifically related to its specialist function (*Candetti Constructions Pty Ltd v Fonteyn*, 2012, p. 250).

However, Parsons J, with whom Farrell J agreed, rejected this. His opinion found that Candetti's responsibility for site safety included control of the relevant opening, and that it was obliged to put in place the reasonably practicable measures of providing secure fencing and a safe system of work (which was put in place after the accident) to ensure workers were safe from the risk of falling (*Candetti Constructions Pty Ltd v Fonteyn*, 2012, p. 288).

The following factors were relevant to establishing that Candetti should have issued directions to the specialist contractor: Candetti's policies dealt with working from heights and fall protection; its managers had regularly inspected the site and directed workers to replace, or had themselves replaced, fencing that had been removed; Candetti had general responsibility for the site; other trades were using the opening and Candetti was responsible for scheduling and coordinating work; the lack of an adequate system of fall protection should have been apparent to Candetti; the potential hazard from the unguarded opening applied to its own employees and other workers; and it should have been apparent that Candetti's system for managing and supervising fall protection and its inductions on this issue were inadequate, given workers were constantly removing the bunting and failing to replace it (*Candetti Constructions Pty Ltd v Fonteyn*, 2012, pp. 283, 287).

The above cases show that parties will be held accountable for a failure to supervise, issue directions to, or take other reasonably practicable steps in relation to, specialist contractors, where the hazard is *within* the scope of the party's control, knowledge or expertise or is obvious. This is particularly so where they have taken such measures for their own employees, or others in the past. In these circumstances, parties will not be able to simply rely on the specialist contractor to take care of all safety matters associated with its work, and must remain proactive and vigilant in relation to those matters it is reasonably practicable for it to deal with.

However, even in this context, the issue of whether or not it is reasonably practicable to *exercise* control or issue directions in respect of a matter, which is within the party's *capacity* to control and its expertise, is a question of fact, requiring evidence. So much is clear from the decision in *Baiada Poultry Pty Ltd v The Queen* (2012), in which the High Court reversed the majority of the Victorian Court of Appeal's decision that it was 'entirely practicable for [Baiada] to require the contractors to put loading and unloading safety measures in place and to check whether those safety measures were being observed from time to time' (*Baiada Poultry Pty Ltd v The Queen*, 2011; cited in *Baiada Poultry Pty Ltd v The Queen*, 2012, [19]).

Baiada Poultry Pty Ltd (Baiada) carried on a poultry processing business. It contracted with growers, chicken catchers and transport operators to grow, round up, and transport, respectively, the chickens to its processing plant. In the process of loading some remaining empty chicken crates onto a trailer (once the chicken catchers had run out of chickens), an unlicensed forklift operator, whose supervisor had walked away to make a phone call, got stuck. The truck driver went to assist, and as the forklift operator moved the crates, they toppled onto the truck driver, killing him.

The evidence established that Baiada had the capacity to issue safety directions to its contractors and that it had in place at its own plant traffic management policies, which would have eliminated or reduced the hazard. However, Baiada had argued that a practicable method for Baiada to discharge its duty was to rely on the experience of its subcontractors and that it was 'entitled' to 'rely' on its expert contractors. As the majority noted, this argument did not engage with the words of the legislation, and the effect of the contention was that it had not been established that it was reasonably practicable for Baiada to exercise its right to control its subcontractors' activities (*Baiada Poultry Pty Ltd v The Queen*, 2012, [16]).

The majority stated that the duty does not require the duty holder to take every *possible* step that could be taken, or even a step that could have been taken that might have impacted on safety (*Baiada Poultry Pty Ltd v The Queen*, 2012, [15]). Rather, the task remains one of assessing whether the duty holder has discharged its duty, so far as is reasonably practicable (*Baiada Poultry Pty Ltd v The Queen*, 2012, [16]). The majority also stated that having a capacity to exercise control does not necessarily mean it is reasonably practicable to exercise it, rather this only demonstrates a step is possible (*Baiada Poultry Pty Ltd v The Queen*, 2012, [33]). Given the above factual circumstances and the other cases referred to above, it is difficult to see how it could not have been reasonably practicable for Baiada to exercise its capacity to control traffic management procedures of its subcontractors, particularly as it had done so in the past with another contractor (*Baiada Poultry Pty Ltd v The Queen*, 2011, [27]). However, this was a case involving the proviso (i.e., the question of whether, not withstanding that the trial judge made an error in directing the jury on this point the guilty verdict be upheld, as there was no miscarriage of justice), and it is understandable that the High Court was cautious in reaching a conclusion that there was no reasonable inference consistent with Baiada's innocence (*Baiada Poultry Pty Ltd v The Queen*, 2011, [36]).

Justice Heydon's judgment was broadly consistent with the majority's view. However, with due respect, I do not agree with Heydon J's comments that a duty holder need not provide the safest working environment that is reasonably practicable, but rather can provide one amongst a range of practicable environments (*Baiada Poultry Pty Ltd v The Queen*, 2011, [63]), and that even if it was reasonably practicable for Baiada to exercise control, it did not follow that Section 21(1) of the pre-model legislation required this (*Baiada Poultry Pty Ltd v The Queen*, 2011, [70]). This seems inconsistent with the approach that parties should take proactive steps to comply with their safety duties. It is also inconsistent with the approach of a hierarchy of controls.

However, as the next few cases demonstrate, it is less likely that there will be reasonably practicable measures for a party to take in the context of matters outside its control, knowledge or expertise. Nevertheless, it may still be reasonably practicable to call on further experts in certain situations.

Matters outside duty holder's control, knowledge or expertise

Notwithstanding that a party may not have knowledge or expertise in relation to a potential hazard, where the party has notice of the relevant risk, and to address the risk, it needs to draw on expertise it does not have, the Hamersley Iron case (*Hamersley Iron Pty Ltd v Robertson*, 1998, pp. 19–21) stands for the proposition that the party must call on that expertise. In that case, Steytler J was satisfied that it was reasonably practicable for Hamersley Iron Pty Ltd (Hamersley Iron) to have engaged further experts to assess the structural integrity of one of its reclaimers that collapsed, killing an employee.

The reclaimer was a substantial piece of plant and had been modified a number of times. Hamersley Iron had engaged various experts to conduct analyses of the reclaimer at various times over its life. However, a key factor in this case was that other similar large plants had suffered structural collapse a few years prior. Following these incidents (well publicised in the mining industry), Hamersley Iron had received correspondence from the Acting State Mining Engineer and the District Inspector of Mines urging it to conduct comprehensive assessments and inspections of the reclaimer regarding its structural integrity. Rather than doing so, it relied on previous reports that had been commissioned some time earlier.

Essentially, Hamersley Iron's argument was that it did not itself have the relevant expertise 'either to discover the design fault or to question what was done by those experts engaged by it in the course of modifications to the reclaimer'. It argued that, in those circumstances, there was nothing more reasonably practicable for it to do. However, Steytler J distinguished reliance on past expert contractors from a failure to commission further expert reports, particularly in light of the 'state of the knowledge' about the potential hazard of a significant failure of the reclaimer, the potentially catastrophic consequences and risks that could arise from such a failure, and the fact of other recent failures of similar machines (*Hamersley Iron Pty Ltd v Robertson*, 1998, pp. 22–23, 27). Therefore, the distinguishing factor in this case was that Hamersley Iron was on notice of a potential safety risk with the reclaimer and was therefore obliged to take steps to have it assessed by further experts at that time.

In contrast to the Hamersley Iron case, in the next few cases outlined below, the relevant courts ultimately held that it was not reasonably practicable for the parties to do anything more than rely on the specialist contractor and make the relevant assumptions (Cf *Seneviratne v Qantas Airways Ltd*, 2006).

In the first set of cases, the lack of control of the duty holder and the issue of the 'state of the knowledge' played a key role. The cases involved the prosecutions of the principal contractor of a construction project, Devcon Australia Pty Ltd (Devcon), and Tobiassen, the self-employed, registered builder whom Devcon had engaged to manage work on site. Devcon had also engaged a specialist contractor, Kefo Steel Erection Pty Ltd (Kefo), whose director and supervisor on site (Kelsh) was killed when the steel rafter on which he was sitting prior to the accident moved laterally and fell to the floor. The steel rafter had been connected to opposite facing concrete wall panels. Kelsh was hit by a panel when he fell to the floor. The accident arose from a failure to ensure adequate lateral restraint to the steel rafters, and it was this failure that WorkSafe alleged Devcon and Tobiassen should have remedied.

The Devcon case turned on Devcon's lack of control, knowledge and expertise in relation to the 'matter' (lack of adequate restraint), and it was not to the point that Devcon had general control over the workplace (*Reilly v Devcon Australia Pty Ltd*, 2008, pp. 501–503). This put the 'matter' outside the scope of Devcon's duty (*Reilly v Devcon Australia Pty Ltd*, 2008). Similarly, in the Tobiassen prosecutions, the hazard was found to be outside the scope of Tobiassen's 'work' as a self-employed duty holder (the test being different to an employer) (*Reilly v Devcon Australia Pty Ltd*, 2008). Whilst Tobiassen had overall management and supervision of the workplace, he did not have control over every aspect of work on site, nor was he engaged in the specific work for which the specialist contractors were engaged, as these fell outside his expertise (*Reilly v Devcon Australia Pty Ltd*, 2008).

Similarly, in *Compete Scaffold Services Pty Ltd v Adelaide Brighton Cement Ltd* (2001) (Complete Scaffolding), although Adelaide Brighton Cement Pty Ltd (ABC) had overall control of the workplace, including a system of permits for confined spaces and general site inductions and procedures, it was not found to have control over the hazard (unstable scaffolding), which was the task that it had hired the specialist scaffold contractor to do (*Compete Scaffold Services Pty Ltd v Adelaide Brighton Cement Ltd*, 2001, [55]–[56]). This is also similar to the common law case of *Leighton Contractors Pty Ltd v Fox* (2009), where the principal contractor was found not to be under a duty to include in its general site induction matters relating to the detail of the specialist contractors' work (*Leighton Contractors Pty Ltd v Fox*, 2009). However, Candetti's case referred to above was different, in that the court found that guarding the opening was not a matter of detail of the specialist contractor's work (*Candetti Constructions Pty Ltd v Fonteyn*, 2012).

It will often not be reasonably practicable for the party to *itself* issue directions or supervise the specialist contractor's work, because of its lack of control, knowledge or expertise regarding the relevant hazard or matter (*Laing O'Rourke (BMC) Pty Ltd v Kirwin*, 2011). As Stuart-Smith LJ said in *R v Associated Octel Ltd* (1994, p. 1062):

> ... the question of control may be very relevant to what is reasonably practicable. In most cases the employer/principal has no control over how a competent or expert contractor does the work. It is one of the reasons why [it] employs such a person—that [the person] has the skill and expertise, including

> knowledge of appropriate safety precautions which [the duty holder itself] may not have. [The duty holder] may be entitled to rely on the contractor to see that the work is carried out safely, both so far as the contractor's [workers] are concerned and others, including [the duty holder's] own employees and members of the public; and [the duty holder] cannot be expected to supervise them to see that they are applying the necessary safety precautions. It may not be reasonably practicable for [the duty holder] to do other than rely on the independent contractor.

Similarly, in Devcon's appeal (*Reilly v Devcon Australia Pty Ltd*, 2008, p. 503) the Court of Appeal noted:

> There is no real scope for a principal (lacking the requisite expertise) to exercise actual control over the detailed manner of performance of work by a specialist subcontractor. If it endeavoured to do so, this would be more likely to lead to hazards than to avoid them ...
>
> (See also Compete Scaffold Services Pty Ltd v Adelaide Brighton Cement Ltd, 2001, [56]–[57]).

However, given its lack of expertise and control, should Devcon have engaged further experts to supervise the specialist contractor it had already engaged? The Court of Appeal (*Reilly v Devcon Australia Pty Ltd*, 2008, p. 503) also did not think so. I stated:

> [WorkSafe] suggested ... that the principal might be required, in such a case, to engage an expert to oversee the method of work adopted by the expert subcontractor. That solution seems to us to be unworkable. A builder (for example) would have to 'double up', at significant cost, on contractors having specialist expertise. Work performed by a plumber or electrician would have to be overseen by another plumber or electrician (whose manner of supervision of the work of the first plumber or electrician would, on this construction, also be subject of the control of the builder).

The factors that were relevant to the finding that there was nothing else reasonably practicable for Devcon to do other than rely on its specialist contractor included that the hazard and control measures arose out of the specialist work and fell within the scope of the specialist contractor's (and outside Devcon's) expertise, and that the hazard was not otherwise apparent (*Reilly v Devcon Australia Pty Ltd*, 2008, pp. 492, 506, 507, 509). This formed part of the objective assessment of the state of knowledge factor in the reasonably practicable definition (*Reilly v Devcon Australia Pty Ltd*, 2008, p. 509). Unlike in the Hamersley Iron case, in the Devcon and Tobiassen cases, there was no notice of the potential hazard, nor any indication that Kefo was performing the work other than carefully and safely (*Reilly v Devcon Australia Pty Ltd*, 2008, p. 503). As the Court of Appeal said in Tobiassen's appeal (*Tobiassen v Reilly*, 2009, p. 228):

> Prior to the accident, there was nothing to indicate that Kefo lacked the relevant expertise or that it may not carry out, or was not carrying out, the work in a safe manner. Nor was there then anything to suggest to the appellant (or a reasonable builder in his position) that additional expert advice was required in order to carry out the work safely. In the circumstances, it is not apparent what other reasonably practical steps the appellant could have taken to prevent the accident.

Although the means of addressing the hazard was known within the wider construction industry, unlike in the Hamersley Iron case, Tobiassen did not breach his duty by failing to call on further experts (structural engineers) to supervise the expert rigging company. The Court of Appeal were not persuaded that the 'state of knowledge' factor included that the risk was known within the industry, but not known, nor reasonably expected to be known, by the duty holder (*Tobiassen v Reilly*, 2009, pp. 225–227). Whilst this may not appear consistent with a proactive approach, it is also easy, with the benefit of hindsight, to say that something more should have been done.

This was what Hall J said in respect of the prosecution of the Fortescue Metals Group (FMG) and its subsidiary, The Pilbara Infrastructure Pty Ltd (TPI), following the death and injuries to workers arising out of the destruction of temporary accommodation units (*dongas*) at a remote camp in the Pilbara when it was hit by Tropical Cyclone George. The workers housed in the camp were construction contractors, including employees and subcontractors of Laing O'Rourke (BMC) Pty Ltd (Laing), working on FMG's railway linking its mine to port facilities in Port Hedland.

FMG and TPI had engaged expert contractors to supply and install the dongas, including Spotless to prepare the request for tender (RFT), Spunbrood Pty Ltd trading as NT Link (NT Link) to supply and install the dongas, and Mr Lawry, who was working for WorleyParsons, to supervise the installation of the dongas. Additionally, the local Shire approved the relevant designs and issued a building licence. However, throughout the entire process, the wrong wind specification had been used, which resulted in the dongas not being built to withstand the potential wind forces for the region, which was notorious for cyclones between November and March each year. Further, despite supervision by Mr Lawry of NT Link's installation, the welds of the dongas to their tie downs (which held them to the concrete footings) were poor and performed by an unqualified welder.

FMG and TPI, and NT Link, had each assumed that the other party would ensure that the design specifications were correct, and that because the Shire approved the building plans, they were correct. Laing had also made assumptions about the suitability of the dongas to be used as safe refuge in the event of a cyclone, including by relying on the expert contractors engaged by FMG and TPI, and the Shire approval. FMG and TPI had relied on the expert contractors to build dongas that were compliant to relevant safety standards. They argued there was nothing more that they could do that was reasonably practicable.

In relation to an argument that further engineering experts should have been engaged to review the donga wind design specifications, Justice Hall said (*Kirwin v The Pilbara Infrastructure Pty Ltd*, 2012, [147]):

> It is always possible to imagine a further step, an additional check or a second opinion that could be obtained, particularly with the benefit of hindsight. ... The question is not whether something else could conceivably be done, but whether it was reasonably practicable to expect principals in the position of [FMG and TPI] to do more.

In the context of what it was reasonably practicable for FMG and TPI to do, Hall J considered matters going to the 'state of knowledge of FMG and TPI. The key factors in relation to this issue were that neither FMG nor TPI had relevant expertise to themselves supervise or check the work of their experts or know that the specifications were wrong (*Kirwin v The Pilbara Infrastructure Pty Ltd*, 2012, [119]–[121]); they had engaged apparently competent experts, including engineers in WorleyParsons (*Kirwin v The Pilbara Infrastructure Pty Ltd*, 2012, [125]–[126], [156]); they had relied on representations that NT Link would obtain engineering advice (which it did not, unbeknownst to FMG and TPI) (*Kirwin v The Pilbara Infrastructure Pty Ltd*, 2012, [125]–[126], [156]); the contractual documentation placed the obligation on NT Link to ensure the specifications were correct (*Kirwin v The Pilbara Infrastructure Pty Ltd*, 2012, [114]); there was no notice of any problem with the specifications or method of work, as all documents were consistently wrong (*Kirwin v The Pilbara Infrastructure Pty Ltd*, 2012, [137]–[138]), neither the Shire (*Kirwin v The Pilbara Infrastructure Pty Ltd*, 2012, [137], [174]), nor NT Link challenged the specifications (although the director of NT Link checked the specifications and had reason to doubt them, he remained silent and this was relevant) (*Kirwin v The Pilbara Infrastructure Pty Ltd*, 2012, [136]); NT Link's related body corporate retained ownership of the dongas and this reinforced the belief that NT Link would exercise care and not put significant assets at risk (*Kirwin v The Pilbara Infrastructure Pty Ltd*, 2012, [137], [141]); and Mr Lawry did not raise with FMG/TPI any issues regarding the installation he was supervising. It is important to note Hall J's observation that had FMG and TPI been alerted to the issue, 'it might well have been that there were other reasonably practicable steps that they could take' (*Kirwin v The Pilbara Infrastructure Pty Ltd*, 2012, [136]).

It is clear that each of FMG and TPI, as well as Laing, made certain assumptions in relation to the suitability of the dongas as safe refuge in the event of a cyclone. The reasonableness of these assumptions was considered by Hall J as another matter relevant to the 'state of knowledge' of the parties (*Kirwin v The Pilbara Infrastructure Pty Ltd*, 2012, [150]). His Honour said that these assumptions demonstrated reliance on experts, rather than a neglect of the duty (*Kirwin v The Pilbara Infrastructure Pty Ltd*, 2012, [150]). Earlier in his judgment, His Honour referred to the Laing appeal and stated (*Kirwin v The Pilbara Infrastructure Pty Ltd*, 2012, [108]):

> It is useful to note in this context that the fact of making assumptions is not in itself inappropriate; it is the circumstances in which any such assumptions are made that may be relevant in determining whether an employer or principal has done all that is reasonably practicable in the circumstances. It is unlikely to be enough for a person to merely assume that someone else will attend to safety requirements, but if such an assumption is based upon inquiries made, assurances given, a reasonable belief as to the skills of those responsible for construction and a reasonable belief that regulatory approval has been obtained for the buildings, it may be well-founded.

In the first Laing appeal, Murray J had found that although Laing's assumptions may have seemed reasonable to Laing, it had not taken all reasonably practicable steps to alleviate the hazard and ensure the workers had safe refuge (*Kirwin v Laing O'Rourke* [*BMC*] *Pty Ltd*, 2010,

[80]–[83]). The steps which Murray J concluded were reasonably practicable were to conduct an inspection, make enquiries, or seek expert advice to check that its assumption was justified (*Kirwin v Laing O'Rourke (BMC) Pty Ltd*, 2010, [78]–[83]). However, this was rejected by the Court of Appeal. Chief Justice Wayne Martin of the Court of Appeal noted that, taking the argument to its logical conclusion, all duty holders whose workers were placed in accommodations in cyclonic regions would be required to take similar steps, including obtaining unspecified engineering advice in relation to the accommodation, to satisfy their safety duties (*Laing O'Rourke (BMC) Pty Ltd v Kirwin*, 2011, [4]–[5]). The Chief Justice stated this was 'plainly impracticable' (*Laing O'Rourke* [BMC] *Pty Ltd v Kirwin*, 2011, [4]–[5]). Murphy JA emphasised there was nothing to suggest to Laing that the Shire-approved dongas were unsuitable refuge, in the context where dongas were commonly used as shelters in work environments in cyclone affected areas (*Laing O'Rourke (BMC) Pty Ltd v Kirwin*, 2011, [68]–[69]). In this regard, the judge did not consider that the risk was foreseeable, and indeed, the expert evidence suggested reliance on Shire approval was reasonable (*Laing O'Rourke (BMC) Pty Ltd v Kirwin*, 2011, [49], [68]).

This raises the question of whether parties should take precautionary steps on the basis that expert contractors may, like employees, make errors which may potentially impact on the safety of workers, as occurred in the Cyclone George disaster. WorkSafe argued that a proactive approach needed to be taken, and duty holders should guard against potential errors by the contractors it had engaged *Kirwin v The Pilbara Infrastructure Pty Ltd*, 2012, [175]–[182]; *Laing O'Rourke (BMC) Pty Ltd v Kirwin*, 2011, [42]).

However, Murphy JA in the Laing appeal and Hall J in the FMG/TPI appeal both considered that the errors in the assumptions only emerged with the benefit of hindsight (*Laing O'Rourke* [BMC] *Pty Ltd v Kirwin*, 2011, [69]; *Kirwin v The Pilbara Infrastructure Pty Ltd*, 2012, [181]–[182]).

Finally, in this situation, NT Link also made assumptions about the accuracy of the specifications in the RFT, the Shire approval, and when its director independently did a 'crude calculation', he took a 'near enough is good enough approach' (*Kirwin v The Pilbara Infrastructure Pty Ltd*, 2012, [130]–[136]). Unfortunately, this meant that no one picked up the incorrect specification, which led to the tragic loss of life and injury during Cyclone George.

The above cases therefore show that there are often gaps in the context of engaging expert contractors. These gaps may occur because of the different knowledge or control over the specialist work activity; a resultant lack of notice or awareness of safety issues regarding the expert contractor's method of work; assumptions being made or reliance being placed on the specialist contractor; a failure to expressly advise other parties of any assumption or reliance; and parties leaving contractors to perform their tasks rather than being proactive, even in relation to matters where they have control. It is in this context that the new express duty in the model legislation to engage in consultation, coordination and cooperation with other duty holders becomes significant.

Duty to consult, coordinate and cooperate

Consultation with workers and their representatives has been recognised as an important feature of safety laws, at least since the Robens Report expressly addressed this issue in the 1960s (Dunn and Chennell, 2012). Additionally, a number of cases have considered the issues and hazards arising from the co-ordination of the activities of various workers, giving rise to duties on principal contractors (Tooma 2012; *Stevens v Brodribb Sawmilling Co Pty Ltd,* 1986; *Candetti Constructions Pty Ltd v Fonteyn,* 2012). Further, the *ILO Convention 155* (1981), Article 17, which provides a framework for international regulation of workplace safety and health matters, requires duty holders engaging in activities simultaneously at one workplace to 'collaborate in applying the requirements of the Convention'. This is known as 'horizontal consultation' Note this kind of duty was not recommended in the mining safety context, instead, preferring a vertical management system: see Queensland Government 2012. (Johnstone and Tooma 2012). However, whilst such horizontal consultation and coordination may have been implied in the common law and pre-model legislation duties in the past (see, for example, *Mainbrace Constructions Pty Ltd v WorkCover Authority of New South Wales* [*Inspector Charles*], 2000), under Section 46 of the model legislation this issue has, for the first time in Australia, been given express statutory recognition as an important part of the process of achieving safety in the workplace.

The background to the duty is found in the review panel's two reports. It recommended this duty be expressly legislated as part of the key common features of all duties of care, including, specifically, the concurrent, non-delegable nature of the duties (Stewart-Crompton, Mayman and Sherriff, 2008). The review panel also observed that the duties limited by 'control' in the existing pre-model framework had produced some confusion and uncertainty about their scope, which in turn had resulted in gaps in the provision of health and safety protection where duty holders believed that others were providing for safety protection, but in fact no one was; had created an inefficient use of limited resources through duplication; and, in some cases, had led to duty holders attempting to pass on control rather than focussing on safety (Stewart-Crompton, Mayman and Sherriff, 2008. Sherriff, 2007). The review panel considered that one of the ways these issues could be addressed was by introducing a duty to effectively co-ordinate activities between duty holders (Stewart-Crompton, Mayman and Sherriff, 2008).

The rationale for the new duty was set out further in the review panel's second report. It stated that a PCBU 'may not have a full understanding of the finer detail or subtleties of the work or working conditions' (Stewart-Crompton, Mayman and Sherriff, 2009). A similar rational is found in the Explanatory Memorandum (Safe Work Australia 2010, [194]), which explains:

> Managing work health and safety risks is more effective if duty holders exchange information on how the work should be done so that it is without risk to health and safety. Co-operating with other duty holders and co-operating activities is particularly important for workplaces where there are multiple PCBUs.

Safe Work Australia and various other state regulators have released a Code of Practice to assist duty holders to meet their obligations in relation to this duty (Safe Work Australia 2011a). It emphasises that the horizontal duty will assist in addressing gaps in managing health and safety risks that might occur from a lack of understanding of how activities may add to hazards and risks to which workers may be exposed; from assumptions being made by duty holders regarding who is taking care of a particular matter; and where the person who takes action may not be the best person to do so.

The duty requires duty holders to, so far as is reasonably practicable, consult, co-operate and co-ordinate activities with all other duty holders who have a duty in relation to the same matter as it does. Like other statutory, safety duties, this duty is also limited by the concept of 'reasonably practicable', but the ordinary meaning of that phrase, not the statutory definition, applies in this context (Safe Work Australia, 2010. Tooma, 2012). Additionally, like the primary duty, the duty in Section 46 of the model legislation is a continuous duty and must be engaged in at the multiple stages of the contracting relationship referred to previously, not as a 'once off' exercise (Safe Work Australia, 2011a). Further, it is not sufficient to simply consult with other duty holders; the three concepts are distinct and all three must be complied with, so far as is reasonably practicable (Tooma, 2012).

In terms of the content of the duty, the three concepts are not defined, and take their ordinary meaning. The Code of Practice provides practical guidance on what is meant by the three concepts. However, the precise nature of the consultation, co-operation and co-ordination in any given case will depend on the party's circumstances and the work context, including the nature and complexity of work arrangements. Consultation includes the sharing of relevant information in relation to the particular matter, resulting in a shared understanding of the nature of risks and control measures to address them. Case law from the industrial context may assist in interpreting what is required to genuinely consult. These cases prove that consultation must not be a 'perfunctory' or 'empty' exercise, or 'mere formality', but rather a 'bona fide opportunity to influence' the outcome or decision (Tooma, 2012; *CEPu v QR Ltd*, 2000; CPSu v Vodatone Network pty, Ltd, 2001; Re *Ivw Enterprises Limited* v Duffy, 1985). Cooperation entails providing assistance to, and not hindering, other duty holders, and implementing any agreed safety arrangements. It also involves not obstructing consultation efforts (Safe Work Australia, 2011a; Dunn and Chennell, 2012; Tooma, 2012). Co-ordination requires parties to plan and organise activities to limit risks and ensure each duty holder can effectively discharge their duties, which may include scheduling, arranging and locating activities, and implementing necessary preconditions before activities begin (Safe Work Australia, 2011a; Dunn and Chennell, 2012; Tooma, 2012). However, whilst the duty encourages co-ordination to minimise duplication and wasted resources, and parties may agree on a particular party with expertise or resources to address a specific risk, the duty holder is still responsible for ensuring, so far as is reasonably practicable, that the particular risk is addressed (Sherriff and Tooma, 2010).

It is clear from the above that the focus needs to be on outcomes, not just the process (Dunn and Chennell, 2012). The obligations are aimed at fostering effective communication and planning, and a shared understanding, in relation to safety matters (Safe Work Australia, 2011a,b). This will, in turn, ensure clarity around what parties are doing in relation risks or control measures, where they are relying on other parties and where assumptions are being made, as well as the impact that their activities may have on others, and other parties' activities may have on them.

In situations involving parties engaging expert contractors, there is a prospect that the dialogue engaged in with other duty holders about safety matters, including the expert contractor, will facilitate the identification of any gaps, or draw attention to any safety risks regarding the expert contractor's method of work. This would then put the duty holders on notice of these safety risks or gaps and oblige them to further discuss the risk to determine reasonably practicable methods of eliminating or mitigating it, which may include engaging further experts. The express duty in the model legislation puts the overlapping, concurrent nature of the duties and its potential impact on safety at the forefront of parties' minds.

Of course, there may still be gaps that are not identified in this process. The duty is not a panacea, and parties will always need to be proactive, creative and flexible in their approach. Further, in practice, there may be difficulties arising from differing points of view and disputes about the appropriate outcome, and principal contractors wishing to exercise control in directing work arrangements. Such issues should be dealt with in the usual dispute resolution processes (Master Builders Association, 2011). However, if this duty is built into an overall safety management structure, this will assist parties in meeting their duties.

Effective contractor safety management requires a holistic, systematic, practical approach with a commitment to safety and innovative ways of achieving that end. As Tooma has pointed out, each of contractor, client or supplier safety management requires parties to enquire into safety-related matters at all stages of the party's involvement with the contractor, client or supplier, and adopt a systematic, 'whole relationship' approach (Tooma, 2011). This includes engaging proactively on safety matters from the selection process, to risk assessment, documentation, delivery and performance, and assessment. A similar approach has been propounded by Inns, arguing that a systematic, but flexible, approach is required to properly and effectively manage the unique challenges faced by parties in the context of engaging contractors (Dunn and Chennell, 2012). Further, McCartney (2012) has identified some useful considerations in managing expert safety issues, including exercising due diligence in assessing and appointing contractors, identifying reliance on contractors, ensuring regular reporting, auditing and verification on compliance, including by way of safety key performance indicators, complying with the Section 46 duties and acting immediately and proactively in relation to identified breaches, incidents or unsafe practices. Whilst this duty has not received much attention to date, it has the potential to improve safety outcomes if utilised effectively as an integral part of each aspect of a holistic approach to safety issues (Master Builders Association, 2011).

Conclusion

In this chapter, I have discussed some of the key issues arising in the context of parties engaging specialist contractors. I have demonstrated that this is a complex area, as is often the case in contemporary work environments, involving multiple duty holders, who will often owe duties in various capacities to different persons.

I have argued that this complexity can give rise to gaps in relation to safety matters in the context of parties engaging specialist contractors—gaps in relation to assumptions that another party may be taking care of a safety issue, gaps in coordinating tasks, where hazards may arise from the interaction of different work activities and gaps in relation to knowledge of potential hazards and controls regarding specialist work for which specialist contractors are engaged and when further experts may need to be called to assist.

In this context, the new, express duty in the model legislation for duty holders to engage in horizontal consultation, co-ordination and co-operation is especially significant. It has the potential, if properly utilised, to increase a dialogue between duty holders and highlight any potential shortcomings in safety managements, or to identify assumptions that may be misplaced. Of course, there may still be cases where such dialogue will not result in a hazard being avoided, but this duty plays an important normative role in encouraging a safety dialogue. It remains to be seen how parties will approach this duty in practice, but it is essential that parties continue to actively manage contractor safety issues, in a two-way approach. To be effective, the communication between parties must be genuine, informed and comprehensive, not just a 'ticking the box' exercise. It is hoped that this duty will have a positive impact on safety outcomes and culture.

References

Baiada Poultry Pty Ltd v R (2012) 286 ALR 421.

Baiada Poultry Pty Ltd v R (2011) 203 IR 396.

Bluff, L., Gunningham, N. and Johnstone, R., (2004). *OHS Regulation for a Changing World of Work*. Sydney: The Federation Press.

Candetti Constructions Pty Ltd v Fonteyn (2012) 213 IR 246.

CEPU v QR Ltd (2010) FCA 591.

CPSU v Vodafone Network Pty Ltd (2001) AIRC 1189

Chugg v Pacific Dunlop Ltd (1990) 170 CLR 249.

Complete Scaffold Services Pty Ltd v Adelaide Brighton Cement Ltd (2001) SASC 199.

Creighton, B. and Stewart, A. (eds) (2010). *Labour Law* (5th ed.). Sydney: The Federation Press.

Crimmins v Stevedoring Industry Finance Committee (1999) 200 CLR 1.

Dunn, C. and Chennell, S. (2012). *Australian Master Work Health and Safety Guide*. Sydney: CCH Australia Ltd.

Foster, N. (2010). General risks or specific measures? The High Court decision in Kirk 23. *Australian Labour Law Journal* 230.

Foster, N. (2012). *Workplace Health and Safety Law in Australia*. Sydney: LexisNexis.

Hamersley Iron Pty Ltd v Robertson (Unreported, WASC, Steytler J, 2/10/1998).

Holmes v R E Spence & Co Pty Ltd (1993) 5 VIR 119.

ILO Convention 155. (1981). Occupational Safety and Health Convention.

Johnstone, R. (1999). Paradigm crossed? The statutory occupational health and safety obligations of the business undertaking. *Australian Journal of Labour Law*, 12, 2.

Johnstone, R. and Bluff, E. (2005). The relationship between 'reasonably practicable' and risk management regulation. *Australian Journal of Labour Law*, 18, 197.

Johnstone, R., Bluff, E. and Clayton, A. (2012). *Work Health and Safety Law and Policy*. Sydney: Lawbook Co.

Johnstone, R. and Tooma, M. (2012) *Work Health and Safety Regulation in Australia: The Model Act*. Sydney: The Federation Press.

Kirwin v Laing O'Rourke (BMC) Pty Ltd (2010) WASC 194.
Kirwin v Pilbara Infrastructure Pty Ltd (2012) WASC 99.
Kondis v State Transport Authority (1984) 154 CLR 672.
Laing O'Rourke (BMC) Pty Ltd v Kirwin (2011) WASCA 117.
Leighton Contractors Pty Ltd v Fox (2009) 240 CLR 1.
Mainbrace Constructions Pty Ltd v WorkCover Authority of New South Wales (Inspector Charles) (2000) 102 IR 84.
McCartney, S. (2012). Can you trust your independent contractor with WH&S obligations? *Lexology* (2 April 2012).
Pacific Steel Construction Pty Ltd v Barahona (2009) NSWCA 406.
Queensland Government, Nationally consistent mine safety legislation: Queensland's proposal for a nationally consistent legislative framework (Queensland, 2012).
R v ACR Roofing Pty Ltd (2004) 11 VR 187.
R v Associated Octel Co Ltd (1994) 4 All ER 1051.
R v Australian Char Pty Ltd (1999) 3 VR 834.
Reeve, B. and McCallum, R. (2011). The scope of employers' responsibilities under Australian Occupational Health and Safety Legislation. *Australian Journal of Labour Law*, 24, 189.
Reilly v Devcon Australia Pty Ltd (2008) 36 WAR 492.
Reilly v Devcon Australia Pty Ltd (2007) WASC 106.
Re Tvw Enterprises Limited v Duffy (1985) FCA 251.
Safe Work Australia. (2011a). *Code of Practice: Work Health and Safety Consultation, Co-Operation and Co-Ordination*. Canberra: Safe Work Australia.
Safe Work Australia. (2010). *Explanatory Memorandum—Model Work Health and Safety Bill*. Canberra: Safe Work Australia.
Safe Work Australia. (2011b). *Interpretive Guidelines—Model Work Health and Safety Act: The Meaning of 'Person Conducting a Business, or Undertaking*. Canberra: Safe Work Australia.
Safe Work Australia. (2011c). *Interpretive Guidelines—Model Work Health and Safety Act: The Meaning of 'Reasonably Practicable'*. Canberra: Safe Work Australia.
Sappideen, C. and Vines, P. (eds.). (2011). *Fleming's The Law of Torts* (10th ed.). Sydney: Lawbook Co Australia.
Scott, C. (2005). Extending employers' duties for the workplace safety of contractors. *Australian Journal of Labour Law*, 18, 87.
Seneviratne v Qantas Airways Ltd (2006) NSWIRComm 69.
Sherriff, B. (2007). Occupational Health and Safety: The concept of control in determining responsibilities: A need for clarity. *Australian Business Law Review*, 35, 298.
Sherriff, B. and Tooma, M. (2010). *Understanding the Model Work Health and Safety Act*. Sydney: CCH Australia Ltd.
Slivak v Lurgi (Australia) Pty Ltd (2001) 205 CLR 304.
Stevens v Brodribb Sawmilling Co Pty Ltd (1986) 160 CLR 16.
Stewart-Crompton, R., Mayman, S. and Sherriff, B. (2008). National Review into Model Occupational Health and Safety Laws, First Report to the Workplace Relations Ministers' Council. Canberra: Australian Government.
Stewart-Crompton, R., Mayman, S. and Sherriff, B. (2009). National Review into Model Occupational Health and Safety Laws, Second Report to the Workplace Relations Ministers' Council. Canberra: Australian Government.
Tobiassen v Reilly (2009) 178 IR 213.

Tooma, M. (2012). *Due Diligence: Horizontal and Vertical Consultation*. Sydney: CCH.

Tooma, M. (2011). *Safety, Security, Health and Environment Law* (2nd ed.). Sydney: The Federation Press.

Tooma, M. (2012). *Tooma's Annotated Work Health and Safety Act 2011*. Sydney: Lawbook Company.

chapter four

Four essential elements for saving lives in contractor management

CityCenter and Cosmopolitan projects case study

Sarina M. Maneotis
Sentis

Contents

Typically, when a large-scale tragedy occurs, communities pause to consider what is important, companies re-evaluate their policies and procedures, and ideally changes are made to prevent these incidents from ever happening again. However, workplace fatalities occur frequently within the construction industry and generally aren't afforded this same reflection and appraisal.

The CityCenter and Cosmopolitan construction project in Las Vegas, for example, resulted in eight deaths in 1 year, roughly the same that were reported for the entire decade of the 1990s (Berzon, 2008). Although eight deaths is a disaster by any definition, these fatalities went largely unnoticed by the greater U.S. population because they did not accompany dramatic incidents, such as building collapse or an explosion.

As such, the project largely escaped national media scrutiny, and the resulting calls for change that would normally accompany such disasters. Despite the lack of focus on this otherwise conspicuous failing, there are many lessons to be learned from the CityCenter and Cosmopolitan construction project, particularly with regard to contractor management.

Not only were these projects large in scope, but they were extensive in their use of contractors. Contractors were needed to support nearly every aspect of the build and in total over 7,000 workers were employed at CityCenter it is largely contractors. And, contractors who suffered, with eight deaths between the two job sites (Wise, Morris and Berzon, 2013).

Given the tragedies across CityCenter and Cosmopolitan, it's worthwhile to examine what went wrong and how they can be prevented in the future. To organize the analysis of the events that transpired at these two construction sites, I draw on the framework by Törner and Pousette (2009) for excellent contractor (construction, in particular) management. The four essential elements they cite for effective contractor management are:

1. Project characteristics and nature of the work
   a. Complexity and conditions of the build
2. Organization and structures
   a. Planning
   b. Responsibility for safety (inherent in individual roles)
   c. Safety procedures
   d. Time and budgetary resources
3. Collective values, norms, and behaviors
   a. Safety climate and culture
   b. Interaction and cooperation between employees
4. Individual competence and attitudes
   a. Competence (knowledge, skills, experience)
   b. Attitudes and value towards safety

In the remainder of this chapter, I first present the details of the events at CityCenter and Cosmopolitan. Törner and Pousette's framework is applied to assess the failures of the CityCenter and Cosmopolitan projects in effectively managing contractors and ensuring their safety. I close by making recommendations for improving contractor safety management based on the case study and analysis presented.

Case study

Background

The CityCenter and Cosmopolitan projects were collectively one of the largest construction efforts the world has ever seen (Audi, 2008).

Both overseen by the same general contractor, Perini, the CityCenter was budgeted at $8 billion and Cosmopolitan at $3.5 billion. At any given time, there were approximately 7000 employees contracted to work on the CityCenter build, which operated on a 24-hour schedule, with individual shifts ranging from 8 to 10 hours (Knightly, 200b). As can be imagined, a variety of electricians, engineers, glass and metal workers, and many other trades were required to complete the projects (Gittleman et al., 2010). In addition to being extremely large in scope, the project was also on a tight timeline. Starting in June 2006, the CityCenter project was scheduled to be completed in 2009. To sweeten the deal, DubaiWorld, an initial investor and parent company of the projects, offered a $100 million bonus for on-time completion (Audi, 2008).

The scope of the project and intense timeline created a situation which, lacking a strong safety culture, would favor meeting production targets over safety. Research has found that when this is the case, workers are more likely to cut corners on safety to do what it takes to meet production demands, such as forgoing proper PPE, or not following the full safety procedures (McGonagle and Kath, 2010).

Sometimes, cutting corners doesn't yield any negative outcomes. However, this was not the case at CityCenter and Cosmopolitan. On August 9, 2007, the first fatality was incurred by CityCenter. A worker was crushed by the manlift. He had not been trained on how to safely operate the lift and in addition, operators and/or supervisors had reportedly not lubricated the manlift to the manufacturer's specifications (Wise, et al. 2013). Initially, OSHA fined the project $21,000, but this was later withdrawn. As Table 4.1 shows, seven other deaths occurred across the projects in the span of 1 year.

Unfortunately, none of these incidents alone moved the project to re-examine how it was managing its contractors, safety procedures, or review employee training. It wasn't until June 2008, when workers held a 1-day strike for a work-site safety assessment, on-site training, and access to union and safety officials, that a large-scale investigation was conducted (Knightly, 2008a).

What went wrong?

After examining Table 4.1, it is clear that a range of factors contributed to these accidents. Interestingly, Perini official Doug Mure indicated it was impossible to know what caused the accidents: "It wasn't all due to one cause or one type of accident. If we could point to a cause and effect, we would fix that" (Audi, 2008). Mure is right in that there was no single cause. Sure, contractors weren't following procedures, but they might not have been trained on them either. They were also likely to be receiving production pressure from their supervisors and perhaps had trouble

Table 4.1 Fatalities and their causes, occurring between August 2007 and May 2008 at CityCenter and Cosmopolitan

Site	Date	Incident	Potential cause(s)	Result
CityCenter	August 9, 2007	Worker crushed by manlift	Worker not trained properly on locking out manlift Supervisors/operators had not lubricated the lift properly	$21,000 OSHA fine, later withdrawn
CityCenter	October 5, 2007	Ironworker fell to his death through faulty metal decking	Employee did not have his harness on The decking had not been completed and had gaps No guards or netting around the decking	$13,500 OSHA fine, later removed
Cosmopolitan	November 27, 2007	Worker fell to his death when the beam he was working on came loose and fell with him attached to it	Employee harnessed himself to the beam he was working on as there were no other options No netting or temporary floors located below the beam	
Cosmopolitan	January 14, 2008	Safety engineer crushed by corner beam	Kicker that was supposed to support the corner beam had been removed	
CityCenter	February 6, 2008	Two carpenters were crushed by a concrete support form	Support forms not secured properly Additional crane support had been removed Employees not trained in how to properly remove the form	$14,00 OSHA fine, later reduced to $7000
CityCenter	April 26, 2008	Electrician fell to his death		
CityCenter	May 31, 2008	Crane oiler crushed by crane		Worker strike in June 2008

Source: Wise, Z., Morris, C. and Berzon, A. (2013). Construction Deaths. *Las Vegas Sun*. Retrieved from http://www.lasvegassun.com/news/topics/construction-deaths/.

communicating with their colleagues from diverse trades and areas of the country. Applying Törner and Pousette's framework, we can group and assess these factors according to four distinct topics, which will be discussed soon.

What Mure was incorrect about was his ability to do something about the fatalities that occurred on his construction site. Recommendations for improving contractor safety management based on this case study are given at the conclusion of the analysis.

Project conditions

One aspect of the project conditions that likely led design, including the fatalities was the overall complexity of the building design, and size of the project. As a result of these complexity factors, a large number and variety of contractors (iron, glass, engineering, etc.) were needed for the project. Collectively, these demanding conditions likely led to gaps in communication between and amongst contractor groups, rushing to allow for the next piece of the job to start; even overstimulation due to the extreme amount of events happening on site can lead to cognitive overload and work errors (Laxmisan et al., 2007).

In support of this interpretation, site inspections in 2009 confirm that the less-than-ideal conditions of the site led to flaws in the building. After inspection, it was seen that the rebars, or supports meant to stabilize the building, were installed incorrectly (Berzon, 2009). The inspector, Ron Lynn indicated that initial inspections should have taken place long ago and that the errors in the systems were not even consistent. Some were simply installed incorrectly and some were in the wrong area. Collectively, this suggests that the complexity of the project created a workspace where these sorts of extreme errors could occur without being observed or corrected. In this case, project conditions may have not only allowed fatal situations to arise, but the building had to be reduced from 49 to 28 floors since inspectors weren't confident in its soundness to hold the amount of weight it was meant to (Berzon, 2009)!

Organization and structures

Organization and structures consist of four parts: planning, roles, procedures, and resources. In the case of the CityCenter and Cosmopolitan projects, it is unclear how safety was accounted for in planning for the project. However, there are clear pieces of information that indicate roles were ambiguous, procedures were either unclear or not followed, and resources may have been less than optimal.

With regard to safety, not only should individuals know the boundaries of their own role, but responsibility for safe behavior and enforcing safe behavior should be clear. The fatalities at CityCenter and Cosmopolitan show that the delineation of responsibilities for safety was at best indistinct. For example, in the first fatality on August 7, 2009, the manlift worker had not been trained on how to lock it out properly. In addition, though, supervisors and operations responsible for lubricating the manlift had failed to do so. Collectively, it appears that roles were not clearly defined, especially in regard to safety, on the projects.

In addition to role ambiguity, there appears to be uncertainty surrounding safety procedures. It appears that procedures were missing, unclear, or clearly written but not enforced. In any event, safety procedures are most effective when clearly documented, widely available, and enforced to the extent that they are always followed. As Table 4.1 shows, many of the fatalities seem associated with individuals not following procedure. For example, in the October 5 and November 27 fall, safety netting was absent. In addition, the employees involved in these incidents were either not using or misusing their safety harnesses. Had these procedures been enforced, these incidents may not have occurred.

Finally, resources seemed to be scarce on the work site, which likely led to work-safety tension. That is, due to the tight timeline and the bonus associated with finishing on time (Audi, 2008), it's likely that employees were encouraged to work as quickly as possible. This is in fact supported by reports (Audi, 2008). Unfortunately, when this approach is taken, it is often at the expense of safety (McGonagle and Kath, 2010). Conversely, Törner and Pousette (2009) recommend that allowing extra time for project completion, and ensuring adequate resources for safety equipment (such as harnesses and netting) can help facilitate better productivity–safety balance.

Collective values, norms and behaviors

In addition to the basic work environment and practices, the social fabric of the job site is also important. This can be challenging for most contractor managers (Farr, Walesh and Forsythe, 1997). Bringing together diverse groups of people who work together for only short periods leaves a smaller opportunity to build a mature safety culture and teamwork than with a long-term workforce. It's no surprise, then, that climate and culture, interaction and cooperation—facets of this category—can help explain the incidents at CityCenter and Cosmopolitan.

As part of the negotiations that occurred after the worker strike in June 2008, a multidisciplinary team was commissioned to conduct a safety culture analysis of the CityCenter and Cosmopolitan worksites (Knightly, 2008a). Typically speaking, safety culture is the shared perceptions and beliefs within an organization in regards to safety and its priority (Guldenmund, 2000). To assess the safety culture of the CityCenter and Cosmopolitan workforce, the team conducted interviews, observed the work site, and administered a safety climate survey (safety climate provides a snapshot in time of the safety culture; Zohar, 2010). The project is critical for understanding the working conditions and perceptions of safety at the site as this was the first scientific effort to understand the projects.

In total, 5,268 workers completed the climate assessment, along with 212 supervisors and upper executives (Gittleman et al., 2010). The research also showed the diversity of the workforce: 45% of the workforce was Caucasian, 34% Hispanic, 7% African American, with many other ethnicities represented (Gittleman, 2010). In addition, 11% of the assessments were administered in Spanish. Moreover, a variety of trades were represented (carpenters, electricians, plumbers), and the workforce was drawn from nearly each of the 50 states. Collectively, this diversity makes it challenging to create a unified safety culture. For example, it has been suggested that multicultural teams will have innately different levels of safety knowledge and motivation, and that their different backgrounds pose challenges to communications and interactions that are necessary to build a strong safety culture (Starren, Hornikz and Luijters, 2012). Further, cultural background impacts the effectiveness of safety training initiatives (Burke et al., 2012).

Turning towards the results, its unsurprising that the assessment found that workers viewed the safety culture less positively than did supervisors and upper management, supporting media reports that leaders denied the site was unsafe (Hayes, 2007). In addition, lack of management action (27.8%), health hazards (13.5%), and unsafe procedures (10.2%) were identified as the largest barriers to safety at CityCenter and Cosmopolitan. Interestingly, the overall safety culture at Cosmopolitan was found to be more positive than at CityCenter.

In addition to overall culture, interaction and cooperation with coworkers is also important for highly effective contractor management. Although Törner and Pousette (2009) call this out as separate from safety culture, team-level communication and cooperation is generally assessed under safety culture. Indeed, in Gittleman and colleagues' (2010) work, questions such as, "I assist others to make sure they perform their work safely" were measured. Sixty-nine percent of workers reported that they agreed or strongly agreed they'd help others. This rate was comparable to superintendent- and executive-level employees, but was actually even higher in the foreman (frontline supervisor) group. In addition, 5.8% of respondents indicated that communication was the most significant barrier in achieving high safety standards. Overall, the teamwork aspects of safety culture were comparable.

Individual competence and attitudes

The final category that Törner and Pousette (2009) identify for effective contractor management is focused on the individual worker. Competence, in particular, refers to workers' knowledge skills, and experience. In the case of CityCenter and Cosmopolitan, employee competence is a factor in many of the incidents. For example, the February 6 fatalities resulted in part due to the training on cement form removal not being given. Training is highlighted as an issue in the Gittleman et al. (2010) study where a staggering 70% of employees had reported not receiving OSHA training prior to starting work (part of the strike negotiation actually called for onsite training).

Additionally, fatigue, which has been shown to significantly impact occupational competence (Barnes and Wagner, 2009), was a known issue around the site. Likely spurred by the demanding nature of the project, workers reported to the media that they were working safely but were simply tired (Herman, 2008). Additionally, the safety culture survey undertaken by Gittleman and colleagues (2010) found that 19% of workers 'somewhat' to 'strongly' agreed that fatigue was an issue for them.

Törner and Pousette's 'attitudes' refers to individual values regarding safety, such as priority of safety, personal responsibility for safety, and control over individual safety. Individuals who do not have attitudes that value safety tend to be involved in more incidents (Donald and Canter, 1994). Media reports of the CityCenter and Cosmopolitan work sites indicate that safety attitudes were at the time, lax. In particular, Knightly (2008b) documents instances where workers were spotted at a local bar prior to walking over to the construction site to begin work.

Summary

In conclusion, the incidents stemmed from some combination of shortcomings in safety management in regard to project characteristics, organization and structures, collective values, norms and behaviors, and individual competence and attitudes. Mure was correct: there was no single factor that could be attributed to all of the incidents. However, because all of these components were less than Törner and Pousette's ideal, incidents occurred. Just like Reason's (1990) Swiss cheese model of safety, where precautions fail at multiple levels leading to incidents, the CityCenter and Cosmopolitan projects let the project get out of hand in multiple ways, allowing so many fatal incidents to occur. The point is that the events that transpired support Törner and Pousette's (2009) framework for construction (contractor) management: all components are necessary, and all must be excellent for the success of the project and the safety of the individuals involved.

Changes imperative in contractor management

To help prevent projects from going awry like CityCenter and Cosmopolitan, I offer suggestions targeted at each of the four areas that Törner and Pousette (2009) outline. Drawing on examples from aspects of these areas that were not up to par at CityCenter and Cosmopolitan, as well as safety science research, I suggest that aspects of the environment, leadership, planning, and personnel can all be addressed to promote excellent contractor management. See Table 4.2.

As can be seen in the analysis, complicated project conditions facilitated the fatalities that occurred. Part of the issue with complicated project conditions is that they make communications difficult and tasks can easily be done by the wrong person, or not be completed at all. Accordingly, I suggest promoting clear communication between all levels of workers on site: between supervisors and workers and between contractor groups. Research supports that safety communication is associated with decreased incidents (Hofmann and Morgeson, 1999). Further, keeping the work environment neat and free of debris can also reduce incidents (Swat and Krzychowicz, 1997) and can help reduce the number of distractions in an already complex job site. Finally, checks and balances, especially inspections, should be built into the process to help double-check work processes and procedures. Not having these checks and balances in place is a strong predictor of safety incidents (Bellamy, Geyer and Wilkinson, 2008).

In regards to organization and structures, I suggest that not making safety clearly part of the project plan, inherent in individual roles, and providing and enforcing clear procedures, in addition to a lack of resources, contributed to the fatalities at CityCenter and Cosmopolitan. Accordingly, it is suggested that these aspects be addressed. Specifically, creating a safety mission, goals for safety can be clearly spelled out. When this occurs, the rate of goal achievement increases dramatically (Locke and Latham, 2002). Further, making it clear who is responsible for enforcing safety on site, and what each individual's responsibility for safety is, has been associated with safety culture maturity; mature safety cultures approach safety as a way of life and encounter few incidents (Lawrie, Parker and Hudson, 2006). In addition to having clearly defined safety goals and responsibilities, clarity of safety procedures is also important and is associated with proper safety behaviors and reduced incidents (Cox and Cheyne, 2000).

Beyond the clarity or goals, roles, and procedures, work-safety tension is also important to target. Research shows that when supervisors pressure workers to be productive beyond the employees' normal means, safety is sacrificed (McGonagle and Kath, 2010). Obviously, supervisors should be encouraged to reduce the pressure they place on employee speed, but allowing a contingency time and financial resources can help

Table 4.2 Recommendations for promoting excellent contractor management

Area targeted	Suggestions
Project conditions	Promote clear communication between contractor groups (e.g., electricians and metal workers) and from supervisors downwards. Ensure the environmental hazards are contained. Keep noise and debris to a minimum. Post safety signage to communicate hazardous objects. Build in a system of checks and balances to ensure work is organised and being completed thoroughly.
Organization and structures	Create a safety mission or value statement at the start of the project to clearly spell out the priority and importance for safety. Identify who is responsible for enforcing safe procedures on site. Ensure that person has the knowledge to determine whether work is done safety. Identify the responsibility each individual has for working in a safe manner. Ensure procedures that all contractor groups follow are up to date. Encourage extra procedures if necessary to increase clarity. Avoid pressuring contractors to work quickly to finish. This will only increase errors and incidents. Build in contingency time and budget.
Collective values, norms, and behaviors	Hire leaders and supervisors who will *role model* safety. Leadership often set the tone for the rest of the site's safety culture. Provide consistent onboarding for all contractor groups. Encourage contractor groups to communicate with one another early on to proactively identify where any communication barriers may exist. Administer a short safety climate survey. Ensure all levels and contractor groups have a positive and consistent safety climate. Immediately correct areas where safety climate is less favorable or inconsistent with other groups. Provide easy to use reporting outlets for incidents, near-misses and errors. The more data you collect, the more you can prevent these from happening.
Individual competence and attitudes	Only select contractors who have employees with a high degree of safety knowledge, skills, and experience, as well as positive safety attitudes (these can all be assessed for with selection tests). Provide training to all contractors for consistency. Do not assume the safety training they've had is adequate. Provide booster training throughout the scope of the project to remind contractors of the importance of safety. Conduct safety campaigns to keep safety awareness high throughout the project.

reduce the need to place pressure on employees to avoid going over time and over budget.

To address the collective values, norms, and behaviors holes that existed at CityCenter and Cosmopolitan, we suggest first and foremost ensuring that site management leaders are role modeling positive safety behavior. Research indicates that when leaders have a high safety integrity ('walk the talk' in terms of safety) their teams place a higher priority on safety and also report less work errors (Leroy et al., 2012). An in-depth discussion of safety leadership goes beyond the scope of this chapter, but role modeling has been shown to be one of the most important factors of safety leadership, which corresponds to increased safety climate and decreased incidents (Barling, Loughlin and Kelloway, 2002).

In addition to more prominent and effective safety leadership, consistent onboarding for all contractors can help set the tone for a strong safety climate and ensures everyone has access to the same safety knowledge on site (Batcheller, 2011). Additionally, providing on boarding with a variety of contractors present may encourage early communication between groups, which will be essential later when the job begins. Early after the job begins, a safety climate survey may be administered to 'take the temperature' of the current safety culture. Understanding where issues lie early on when they can be fixed can help prevent major incidents later (Biggs, Dingsdag, Sheahan, Cipolla and Sokolich, 2005). Beyond fixing issues with safety culture, an easy-to-use and accessible reporting system can facilitate learning from errors and near-misses that occur on site; use of these systems is generally increased by a positive safety and error management culture onsite (Evans et al., 2006; Van Dyck, Frese, Baer and Sonnentag, 2005).

Finally, individual competence and attitudes can be targeted through training, selection, and safety campaigns. On the CityCenter and Cosmopolitan projects, it was clear that individual competence was not at its peak due to fatigue (Herman, 2008), safety attitudes were lax (Knightly, 2008b), and employees weren't given the training they needed to perform safety (Gittleman et al., 2010). Accordingly, contractor managers should start by selecting contractors with safety experience, skills, and knowledge. Cost is often a concern, but given the money lost during the CityCenter and Cosmopolitan projects (e.g., OSHA fines, reducing building scope and thus potential revenue due to poor execution), hiring well-qualified personnel is worth the price. After contractors are selected, they should receive thorough safety training before beginning work. Even if they have had training through their own company, don't assume it is adequate. Design a program internally or hire a training firm, but make sure the training is engaging, as training that requires participant interaction is associated with decreased incidents (Burke et al., 2006). Providing booster training or running safety awareness campaigns afterwards can help boost the success of the initial training program (Baldwin and Ford, 1988).

Initiating these actions will help address each factor cited by Törner and Pousette (2009) for contractor management excellence. The time and effort put forth on the front end to ensure safety systems are in place, and personnel are trained and have access to the resources they need can in turn build a positive safety culture. Collectively, these components reduce the risk of tragedy so that disasters such as those at CityCenter and Cosmopolitan do not occur.

References

Audi, T. (2008, June 10). Worker deaths at Las Vegas site spur safety debate. *The Wall Street Journal*. Retrieved from http://online.wsj.com/article/SB121305846017559423.html.

Baldwin, T. T. and Ford, J. K. (1988). Transfer of training: A review and directions for future research. *Personnel Psychology*, *41*(1), 63–105.

Barling, J., Loughlin, C. and Kelloway, E. K. (2002). Development and test of a model linking safety-specific transformational leadership and occupational safety. *Journal of Applied Psychology*, *87*, 488–496.

Barnes, C. M. and Wagner, D. T. (2009). Changing to daylight saving time cuts into sleep and increases workplace injuries. *Journal of Applied Psychology, 94*, 1305–1317.

Batcheller, J. A. (2011). On-boarding and enculturation of new chief nursing officers. *Journal of Nursing Administration, 41*, 235–239.

Bellamy, L. J., Geyer, T. A. W. and Wilkinson, J. (2008). Development of a functional model which integrates human factors, safety management systems and wider organizational issues. *Safety Science, 46*, 461–492.

Berzon, A. (2009, January 8). How did CityCenter tower flaws persist? Failed safeguards puzzle county inspections official. *Las Vegas Sun*. Retrieved from http://www.lasvegassun.com/news/2009/jan/08/how-did-tower-flaws-persist/.

Berzon, A. (2008, April 1). 'Not in this city:' Safety engineer says fundamental change impossible in build-crazy Vegas. *Las Vegas Sun*. Retrieved from http://www.lasvegassun.com/news/2008/apr/01/not-city/.

Biggs, H. C., Dingsdag, D. P., Sheahan, V. L., Cipolla, D. and Sokolich, L. (2005). Utilising a safety culture management approach in the Australian construction industry. *QUT Research Week*, 3–7.

Burke, M. J., Chan-Serafin, S., Salvador, R., Smith, A. and Sarpy, S. A. (2012). The role of national culture and organizational climate in safety training effectiveness. *European Journal of Work and Organizational Psychology, 17*, 133–152.

Burke, M. J., Sarpy, S. A., Smith-Crowe, K., Chan-Serafin, S., Salvador, R. O. and Islam, G. (2006). Relative effectiveness of worker safety and health training methods. *American Journal of Public Health, 96*, 315–324.

Cox, S. J. and Cheyne, A. J. T. (2000). Assessing safety culture in offshore environments. *Safety Science, 34*, 111–129.

Donald, I. and Canter, D. (1994). Employee attitudes and safety in the chemical industry. *Journal of Loss Prevention in the Process Industries*, *7*(3), 203–208.

Evans, S. M., Berry, J. G., Smith, B. J., Esterman, A., Selim, P., O'Shaughnessy, J. and DeWit, M. (2006). Attitudes and barriers to incident reporting: a collaborative hospital study. *Quality and Safety in Health Care, 15*, 39–43.

Farr, J. V., Walesh, S. G. and Forsythe, G. B. (1997). Leadership development for engineering managers. *Journal of Management in Engineering, 13*(4), 38–41.

Gittleman, J. L., Gardner, P. C., Haile, E., Sampson, J. M., Cigularov, K. P. …Chen, P. Y. (2010). [Case Study] CityCenter and Cosmopolitan construction projects, Las Vegas, Nevada: Lessons learned from the use of multiple sources and mixed methods in a safety needs assessment. *Journal of Safety Research, 41*, 263–281.

Guldenmund, F. W. (2000). The nature of safety culture: A review of theory and research. *Safety Science, 34*(1), 215–257.

Hayes, B. (2007, October 6). Iron worker falls to death on Strip. *Las Vegas Review-Journal*. Retrieved from http://www.lvrj.com/news/10284917.html.

Herman, R. (2008, June 14). Six Workers Killed in Construction of Las Vegas "CityCemetery". Retrieved from http://www.wsws.org/en/articles/2008/06/neva-j14.html.

Hofmann, D. A. and Morgeson, F. P. (1999). Safety-related behavior as a social exchange: The role of perceived organizational support and leader–member exchange. *Journal of Applied Psychology, 84*, 286–296.

Knightly, A. M. (2008a, June 3). Workers walk off job. *Las Vegas Review-Journal*. Retrieved from http://www.lvrj.com/news/19483444.html.

Knightly, A. M. (2008b, August 7). Safety issues raised. *Las Vegas Review-Journal*. Retrieved from http://www.lvrj.com/news/26371359.html.

Lawrie, M., Parker, D. and Hudson, P. (2006). Investigating employee perceptions of a framework of safety culture maturity. *Safety Science, 44*, 259–276.

Laxmisan, A., Hakimzada, F., Sayan, O. R., Green, R. A., Zhang, J. and Patel, V. L. (2007). The multitasking clinician: Decision-making and cognitive demand during and after team handoffs in emergency care. *International Journal of Medical Informatics, 76*, 801–811.

Leroy, H., Dierynck, B., Anseel, F., Simons, T., Halbesleben, J. R., McCaughey, D. … Sels, L. (2012). Behavioral integrity for safety, priority of safety, psychological safety, and patient safety: A team-level study. *Journal of Applied Psychology, 97*, 1273–1281.

Locke, E. A. and Latham, G. P. (2002). Building a practically useful theory of goal setting and task motivation: A 35-year odyssey. *American Psychologist, 57*, 705–717.

McGonagle, A. K. and Kath, L. M. (2010). Work-safety tension, perceived risk, and worker injuries: A meso-mediational model. *Journal of Safety Research, 41*, 475–479.

Reason, J. (1990). *Human Error*. Cambridge: Cambridge University Press.

Starren, A. Hornikx, J. and Luijters, K. (2012). Occupational safety in multicultural teams and organizations: A research Agenda. *Safety Science, 52*, 43–49.

Swat, K. and Krzychowicz, G. (1997). Investigation of occupational injuries in a meat processing plant. *Advances in Occupational Ergonomics and Safety*, 595–598.

Törner, M. and Pousette, A. (2009). Safety in construction—A comprehensive description of the characteristics of high safety standards in construction work, from the combined perspectives of supervisors and experienced workers. *Journal of Safety Research, 40*, 399–409.

Van Dyck, C., Frese, M., Baer, M. and Sonnentag, S. (2005). Organizational error management culture and its impact on performance: A two-study replication. *Journal of Applied Psychology 90*, 1228–1240.

Wise, Z., Morris, C. and Berzon, A. (2013). Construction deaths. *Las Vegas Sun*. Retrieved from http://www.lasvegassun.com/news/topics/construction-deaths/.

Zohar, D. (2010). Thirty years of safety climate research: Reflections and future directions. *Accident Analysis and Prevention*, *42*(5), 1517–1522.

chapter five

Contractor safety management for nursing

Janis Jansz, PhD
RN, RM, Dip. Tch., BSc. (Nursing Management).
Grad. Dip. OHS, MPH, PhD
Curtin University, CHURI, School of Business, Edith Cowan University

Contents

Introduction and history

Contractor safety management is an important part of health care management. This research based chapter traces the history of nursing and documents Australia wide laws for nursing practice. It critically examines the duty of care owed by the contractors' host employer, government and health service boards to contract nursing staff (agency nurses), nursing agencies' duty of care, patients' and residents' health and safety responsibilities and the health and safety responsibilities of contract nursing staff. The message in this chapter is that everyone who works in, or who uses, a health service has a responsibility not only to meet their legal duties but to have a mission and a culture of care. Private organisations, in which this culture of care for everyone (including contract staff) who comes on the business premises is achieved, research has demonstrated are profitable, while government organisations achieving this culture of care are more cost effective.

Caring for people who are sick as a vocation has been undertaken for as long as history has been recorded. Elizabeth Fry was concerned about the skills of nursing staff. In 1840 she established the first training school for nurses in Britain in Guy's Hospital (Simkin, 2012a). Nurses were provided with a starting wage of 20 pounds a year and their own uniform. Prior to this, nursing care was performed by both untrained men and women who performed menial tasks related to health care. Nursing was servants' work and not well rewarded financially.

Florence Nightingale was influenced by Elizabeth Fry's view that nurses should be trained, and when she went to the Crimea War in 1854 she took 38 nurses who had been trained by Elizabeth Fry with her. When she arrived at the Scutari hospital in Crimea, Florence Nightingale found "thousands of wounded and ill soldiers were treated in closely packed beds by overworked doctors, male medical orderlies, and untrained women whom she dismissed as drunken and slatternly" (Simkin, 2012b, p. 5). In her first winter 4,077 soldiers died in this hospital, mainly from typhus, typhoid, cholera and dysentery. One of the first tasks that she undertook was to improve ventilation and to organise for the sewers to be cleaned out. "By February 1855, the mortality rate had fallen from 60% to 42.7% and then, once fresh water supply was introduced, it dropped further to 2.2%" (Simkin, 2012b, p. 6).

Florence Nightingale returned to London a heroine. In 1859 she wrote three books: (1) *Notes on Hospital,* (2) *Notes on Nursing: What It Is and What It Is Not,* and (3) *Suggestions for Thought to Searchers after Religious Truths.* In her third book she promoted the removal of restrictions in Britain that prevented women from having a career. With the support of wealthy friends, Florence Nightingale was able to raise 59,000 pounds. In 1860 she used this money to found, at Saint Thomas's Hospital, the Nightingale School and Home for Nurses to allow women who would like to work as a nurse to have a nursing career. Under Florence Nightingale's tuition, nurses were provided with lectures on applying dressings, general hygiene, basic anatomy, and were taught, above all, "how to observe, what symptoms indicate improvement, what evidenced neglect" (Nightingale, 1969, p. 105). Nurses worked under strict discipline. "They could be dismissed for having a 'determined manner', for not wearing a hat, for not acquiescing to a head nurse" (Reverby, 1987, p. 121), for complaining about not having enough to eat (Kalisch and Kalisch, 1975), for questioning hospital rules, or for questioning a doctor's orders (Ashley, 1976). The Nursing

Record 1892 (Tellis, 2003, p. 51) declared nurses to be "the white slaves of hospitals—overworked, underpaid, often more than half-starved inside their walls, or sweated as private outside nurses to produce larger profits for the hospitals, and then, when their health was broken down under the strain, discharged—tossed aside like old worn-out things."

Until the late 1980s in Australia, nursing training was hospital based as it was in Florence Nightingale's days. Nurses were required to live in a nurses' home whilst they completed their education and for as long as they worked at the hospital. "Until 1966 Section 49(2) of the Commonwealth Public Service Act read: *Every female officer shall be deemed to have retired from the Commonwealth service upon her marriage*" (Department of the Prime Minister and Cabinet, 2004, p. 53). The Marriage Bar Act in Australia was also enforced by state and local governments, and private enterprise employers. This Marriage Bar Act affected nurses as they had to retire from work if they married. The reason that the marriage bar was removed was that there was a shortage of labour in industry. It was decided, by politicians in the late 1960s, that women could help to decrease the labour shortage in Australia by working, even if they were married, so the Marriage Bar Act was repealed.

In 1988 a Nursing Career Structure was introduced into Australia. At the same time nursing education in Australia changed from nurses being trained whilst working in hospitals to nurses being educated at university with the Bachelor of Science (Nursing) being the initial qualification. Further nursing qualifications included a graduate certificate, postgraduate diploma, and master's university qualifications in specialities, including nursing management, clinical nursing, midwifery, aged care, palliative care, diabetes, child and adolescent health nursing and for nurse practitioners. With university education, nurses were permitted to live away from their place of work and upon graduation were treated as professional workers who were expected to have the required core body of nursing knowledge. Today, nursing in Australia is a mixture of experienced nurses who have hospital-based training and nurses who have received their nursing education at university.

By 2001, in Australia, it was necessary for all registered nurses to be eligible to register with the Nurses Board in each state, and to have professional indemnity and public liability insurance of $10,000,000 in case they were found guilty of malpractice or negligence that causes injury to a third party, person or property.

Australia-wide laws for safety in nursing practice

By 2012, in Australia, all nurses, including contract agency nurses, were required by the Health Practitioner Regulation National Law to register with the Nursing and Midwifery Board of Australia ("the National Board") to be able to work as a registered nurse or midwife in Australia (Nursing and Midwifery Board of Australia, 2012). Under the Health Practitioner Regulation National Law 2009 this board registers all nurses, midwives, students of nursing and midwifery students. A list of all nurses registered to practice by the Nursing and Midwifery Board of Australia is kept on a public register located at the web address http://www.ahpra.gov.au/Registration/Registers-of-Practitioners.aspx. The Nursing and Midwifery Board of Australia develops professional nursing and midwifery standards, codes of practice, nursing guidelines, position statements, and national registration requirements, and assesses the qualifications of all nurses and midwives who wish to work in Australia.

In Australia the Health Practitioner Regulation National Law is supported by the following laws:

- Health Practitioner Regulation National Law Regulation 2010.
- Queensland Health Practitioner Regulation National Law Act 2009.
- New South Wales Health Practitioner Regulation Act 2009.
- Health Practitioner Regulation National Law (Victoria) Act 2009.
- Australian Capital Territory Health Practitioner Regulation National Law (ACT) Act 2010.
- Northern Territory Health Practitioner Regulation (National Uniform Legislation) Act 2010.
- Health Practitioner Regulation National Law (Tasmania) Act 2010.
- Health Practitioner Regulation National Law (South Australia) Act 2010.
- Health Practitioner Regulation National Law (WA) Act 2010.

The Health Practitioner Regulation National Law Bill 2009 (Australian Government, 2009, pp. 1, 2) documented that the objectives of this national registration and accreditation scheme were:

a. To provide for the protection of the public by ensuring that only health practitioners who are suitably trained and qualified to practice in a competent and ethical manner are registered;
b. To facilitate workforce mobility across Australia by reducing the administrative burden for health practitioners wishing to move between participating jurisdictions or to practice in more than one participating jurisdiction;
c. To facilitate the provision of high-quality education and training of health practitioners;
d. To facilitate the rigorous and responsive assessment of overseas-trained health practitioners;

e. To facilitate access to services provided by health practitioners in accordance with public interest; and
f. To enable the continuous development of a flexible responsive and sustainable Australian health workforce and to enable innovation in the education of, and service delivery by, health practitioners.

Under the Health Practitioner Regulation National Law 2009 clause 31 established the following boards: Medical Board of Australia, Dental Board of Australia, Physiotherapy Board of Australia, Chiropractic Board of Australia, Optometry Board of Australia, Osteopathy Board of Australia, Pharmacy Board of Australia and Nursing and Midwifery Board of Australia. One in every 39 Australians in 2013 was identified as a registered health professional.

The Nursing and Midwifery Board of Australia is supported by the Australian Health Practitioner Regulation Agency (AHPRA), which manages all nursing registrations, renewals and any notification matters. The Australian Nursing and Midwifery Accreditation Council Limited (ANMAC) assesses and provides recommendations for accreditation for all Australian Nursing and Midwifery educational programs of study for endorsement for approval by the Nursing and Midwifery Board of Australia if this program leads to registration with this board. It also monitors and audits Nursing and Midwifery Board of Australia–approved educational programs of study.

Nursing is now a highly regulated profession with all nurses being required to a obtain university degree accredited by the Nursing and Midwifery Board of Australia for their proposed area of practice and to have continual ongoing nursing-related education of at least 20 hours per year, have personal professional indemnity and public liability insurance (in case they make a mistake with patient care) and to pay an annual registration fee to the Australian Health Practitioner Regulation Agency to be allowed to be registered to work as a nurse. These requirements apply to nurses who work as contractors (agency nurses) as well as to those who work in hospitals and nurses who work in the community.

In addition to being registered with the Australian Health Practitioner Regulation Agency contract (agency nurses), nurses who want to work in government hospitals in Western Australia as a registered nurse are also required to pass a compulsory:

- Police clearance;
- Working with Children Check;
- MRSA (Methicillin Resistant Staphylococcus Aureus) Screening;
- Annual Cardio Pulmonary Resuscitation (CPR) competency;
- Annual managing aggressive behaviour course;

- Annual manual handling course;
- Annual fire and emergency management course;
- Nursing/Midwifery Staff Hand Hygiene course;

and to pay a fee to the Health Department of Western Australia to gain permission and to be registered to work in Western Australian government hospitals. These requirements are so that the health care recipient is assured that the registered nurse caring for them is a person of good character with a set of core competencies who will not transmit infection.

Since Florence Nightingale's experience in the Crimea War infection control efforts were an important part of nursing to both protect the health of the patient and of the nurse. The formal education of nurses to be able to work competently as health care professionals that was begun by Elizabeth Fry is now accredited and regulated by the government. Florence Nightingale's opinion that women should legally be allow to have a career as a nurse, whether they are married or not, is now a reality.

Throughout history, whether nurses were untrained servants, hospital-trained nurses or university-educated nurses, there have been nurses who have worked as contractors. Today, many nurses work for nursing agencies. Nursing agencies contract out nurses to work in hospitals, nursing homes and other health care facilities to provide nursing care, to cover sick and other leave for nurses and to make up the staffing numbers when there are more nurses required for patient care than there are nurses available in the host employer's organisation to provide this care. Contract nurses, who are usually called agency nurses, also work in private homes to care for people who require nursing care at home.

Duty of care owed by host employer

Under workplace health and safety laws, for nurses who work as a contractor, a duty of care is owed by both the nursing agency that the nurse works for and by the hospital or other facility that the contractor nurse works in (Jansz and Mills, 2008). In this relationship the place where the nurse goes to work controls the physical work space, equipment required for health care use, work organisation, and the policies, procedures and guidelines. As such the host employer needs to ensure that all of these are safe for the people who will use them. For these factors the duty of care by the host employer is the same for the agency nurse as it is for all nurses who work at this facility (Safety Work Australia, 2012).

In particular the host employer has the responsibility under the Western Australian Occupational Safety and Health 1984 section 19 to, so far as is practicable, provide and maintain a working environment in which the agency nurse is not exposed to hazards. This includes the following:

1. (a) Provide and maintain workplaces, plant, and systems of work such that, so far as is practicable, the employees are not exposed to hazards; and
   (b) Provide such information, instruction, and training to, and supervision of, the employees as is necessary to enable them to perform their work in such a manner that they are not exposed to hazards.
2. In determining the training required to be provided in accordance with subsection (1) (b) regard shall be had to the functions performed by employees and the capacities in which they are employed.

Other states and countries have similar requirements in their workplace health and safety legislation. As part of their duty of care to contract agency nursing staff, the host employer should provide agency nurses with an orientation to the place where they will work their shift.

Provision of information

The host employer should provide the nursing agency manager with specific information about the work situation and any safety factors before the agency nurse is employed to work a shift in a health service area (New South Wales Nurses' Association, 2004). Safe Work Australia (2012, p. 3) documents that the host employer should provide an agency nurse "with a site-specific safety induction outlining work health and safety duties, policies, procedures and practices in the workplace". An agency nurse may only work one 6-hour shift in an area and never come back to work at the same health care facility so these requirements are not always met. Every work shift can mean that the agency nurse has a new workplace layout to learn, has new equipment to work with that the nurse may, or may not have used in other health services before, and that the patients that the nurse cares for are all new to the nurse. Employees new to a workplace are often the most likely to have errors, particularly in health services (Commission of the European Communities, 2002).

The agency nurse usually does not know the clients to be cared for, where the equipment is kept, and where facilities (such as treatment room, pan room, linen storage area) are located. In an unfamiliar workplace, finding where equipment and products are located can take time. Without an orientation, the agency nurse may make errors with patient care because when help is required for information by the agency nurse there is no one available, or willing, to provide this information. When new to a workplace, the agency nurse often does not know the normal work procedures for this workplace as standard workplace procedures vary in different workplaces. Having good facilities, policies, procedures and equipment is a start, but it is important that agency nurses are given an orientation to the area in which they will work so that they can find where the patients' they will care for are located, find the facilities, equipment and products (such as drugs) that they will use and have the normal work procedures that they will need to use during their work shift explained to them as well as where written documentation of workplace procedures and other relevant information are kept. This increases the ability of the agency nurse to work safely and to care more effectively for their allocated patients.

WorkSafe Victoria (2008) provides a case study story of how a private hospital meets its legal requirements to enable agency nurses to work safely at the hospital. As part of the induction, an agency nurse receives an introduction to all hospital emergency procedures and "is given a handy pocket-size guide to emergency codes that can be communicated over the internal telephone system, mobile telephone and security keypads" (p. 18). This orientation would only need to be given once for each hospital for each agency nurse if the procedures and codes do not change. At this private hospital the agency nurse then spends their first working shift with a permanent senior nurse who provides on-the-job information to ensure that the agency nurse understands the risk and risk control strategies used in the area that they are working and the required information that the agency nurse will need to care safely for their patients. This is a practical way for the host employer to meet their duty of care as the agency nurse can ask an experienced staff member any questions to ensure that the work is safe and that they work safely.

Even at the same health service, in some ward or areas an agency nurse may be given an orientation whilst in other areas of the same health service this does not happen. Orientation all depends on who is on duty at the start of the shift rather than on the health service policy and procedures. Even when the agency nurse asks for this orientation the staff on duty may be too busy to provide one. Before they commence work an agency nurse may have to sign that they have had an orientation to the workplace and that they understand all of the work policies and procedures (which they have not seen). Nurses have to sign this document, otherwise they will not be allowed to work the shift. The orientation that an agency nurse signs for may never happen as when the agency nurse arrives at the area for work this nurse does not get shown around the area, or provided with any of the orientation information signed for, but the health service considers that it has signed documentation that it will show in a court of law saying that they have met their legal responsibilities. In healthcare the theory is "If it is documented, it must have been done."

In the above situation the health care service that the agency nurse is sent to is not meeting their duty of care. Although not related to nursing, a similar situation happened at Longford with employee education. In September 1998 the Longford ESSO gas plant in Victoria exploded killing two employees (both were workplace supervisors) and injuring eight other people. During the Royal Commission that followed, ESSO blamed operator error for mixing cold air with warm oil, which ultimately caused the heat exchanger to fracture. At ESSO, operators were given a training module on the dangers of cold temperature embrittlement followed by a written test, the results of which were either a pass, or a need for re-explanation, or coaching required. The training supervisor ticked off the test answers and offered extra training and explanation to those who did not achieve the required results. According to one operator "It took gumption to ask for a re-explanation" (Hopkins, 2000, p. 18). Consequently, some employees wrote correct answers to test questions, but the real concepts they did not understand. The Royal Commission refuted ESSO's claim of "operator error", instead laying the blame on inadequate knowledge due to inadequate education. As a result of this accident at Longford, ESSO was fined $2 million as a penalty for not meeting its duty of care as an employer (Hopkins, 2002).

The following case study describes a situation in which adequate information was not provided to a contract labour-hire workers and which cost the Shoalhaven City Council $753,369 in compensation payments to this worker. In 2008 a Campbell Page labour-hire contract worker was conducting sewerage inspections for maintenance work for the Shoalhaven City Council and was asked by the workplace supervisor to lift a concrete manhole cover that weighed between 75–80 kg. "As he was using a T-bar to lift the cover he injured his right shoulder and spine" (Zenergy Recruitment, 2012, p. 1). On the 23rd of November 2012, the New South Wales court judge decided that as the employer hiring this labour-hire contract employee had not provided the contractor with information about, and training in, safe manual handling practices, and had not provided adequate supervision or conducted an adequate risk assessment of the task, it was responsible for the contractor being injured. Judge Levy identified that the Shoalhaven City Council provided information about, and training in, safe manual handling practices to council employees "but excluded employees of labour-hire companies" (Zenergy Recruitment, 2012, p. 1) from such inductions. The judge refused the council's position that the labour-hire employee had contributory negligence. As well as highlighting the need to provide information, this case study demonstrates the importance of a host employer providing contract labour-hire employees with adequate training to safely use equipment.

Safe use of equipment

As agency nurses work in many different health care services, this nurse often does not know how to use equipment if it is different from equipment in other places that this nurse has worked. For this reason how equipment that the agency nurse will use during the shift actually works should be explained to the agency nurse if she, or he, is unfamiliar with using the equipment. If the agency nurse requests supervision until the person feels competent in using the equipment, this should be provided. It would be appreciated by agency nurses if host employment health services had lectures that agency nurses, who would like to work at the health service, were provided with and training on how to use the host employer's equipment safely. The following story, told by James Reason and cited in Jansz (2011, p. 112), is a good example of why education and supervision in the use of unfamiliar equipment is important.

> *The incident.* During a syringe change-over, a nurse incorrectly re-calibrated a syringe pump delivering a morphine infusion to a patient with stomach cancer, resulting in a fatal overdose.
>
> *The response.* Institution suspended the nurse pending an investigation. She was subsequently given a formal written warning, reinstated and retrained in the use of syringe pumps.
>
> *Incident investigation.* Showed that a Grasby MS26 syringe driver was being used. Whereas this pump is calibrated in mm per hour, a second widely used pump, the Grasby MS16A, is calibrated in mm per day. During the syringe change-over, the nurse applied the calibration principles for the MS16A pump to a MS26 pump.
>
> *Early warning signs.* Two similar errors had recently been reported. Both errors were detected before harm was done.
>
> *Recommendation.* Chief pharmacist and two consultants wrote to management requesting that a single pump be used throughout the Trust.
>
> *Management response.* Suggestion rejected because cost would make it impossible for the institution to stay within the budget limits set by the regional health authority.
>
> *Recurrent system problem.* In all three cases, nurses had been working on understaffed wards. Sisters had complained about not having enough staff to be able to provide a safe level of patient care, but management accepted this as a *"sad fact of life"* and did not act.
>
> *Key situational factors.* Equipment design, workloads, etc., were not thought relevant. Sole focus was on nurses involved: naming, blaming, retraining.

This can happen in the case of agency nurses, too, when they make a mistake due to not being provided with all of the relevant information required.

In 2003, James Reason was appointed Commander of the British Empire for his services in reducing accidents and improving safety in health care services. James Reason developed a model of accident causation that highlights multiple causes of accidents, rather than errors just being the fault of the person who performed the incorrect action. He worked for the Air Force Institute of Aviation Medicine in Britain. In the following model, James Reason identified that accidents were caused by lack of barriers to prevent the risk of hazards causing harm. He looked at accident causes as being more than just an individual's performance and also considered the related latent (hidden) failures that arise from decisions made by upper level workplace managers. (See Figure 5.1.)

In the equipment failure case study, the organisational cause (based on upper management decisions) of this accident was having similar syringe pumps that delivered a drug at different rates. If the chief pharmacist and two consultants' advice had been taken after the near-miss accidents, and only one type of syringe driver was used in the hospital, then this accident probably would not have occurred. If there had been adequate supervision of the nurse, and checking of the syringe driver administration setting by two nurses, then the error probably would have been detected by the second nurse, and this nurse would probably have used the syringe driver correctly. A precondition for this active failure was that the ward that the nurse was working on was understaffed (system failure due to management staff number decision for this ward), and the senior nurses were too busy with patient care to be able to provide the required supervision. Many latent failures are the responsibility

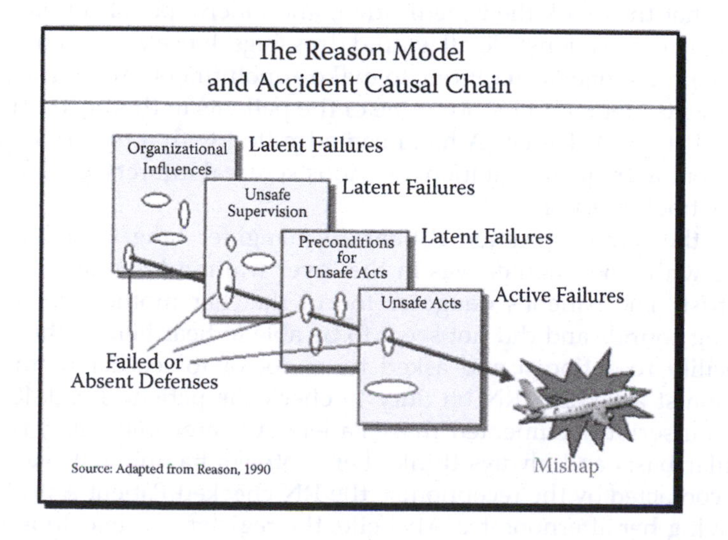

Figure 5.1 Reason's "Swiss cheese" model of human error causation. (From Luxhoj, J. and Kauffeld, K. 2003.)

of the host organisation management who have a responsibility to make the work as safe as possible for agency nurses when they come to work at a host organisation. A factor in this story about the error that this nurse made was work organisation. In many cases, when agency nurses come to work, it is because the area in which they work does not have enough staff for patient care.

Safe systems of work

As well as being unfamiliar with the patients and their care requirements agency nurses can sometime be given more patients to care for than they can provide a safe level of care to. The reason may be due to the number of patients they have to care for, as in the following law case, or due to the level of care required for postoperative or very ill patients as is described in the second law case.

Both law cases come from the Nursing and Midwifery Tribunal of New South Wales (2011). The first case is HCCC v Bottle, 2011. In this case Michelle Bottle was employed to work as a registered nurse (RN) at Berkeley Vale Aged Care Facility. On the day in question Michelle had 46 patients to care for in this aged care facility and was so busy that she did not have time to read each patient's notes from the previous day and shift worked before she wrote her own patient care notes. She was the only RN on duty in this ward and was responsible for giving all medications to the 46 patients (this involved at least three medication rounds, which could take most of the shift time as each patient had to be supervised to ensure that they took their medication, and elderly people often have to take many medications), do all wound dressings, liaise with each patient's medical practitioner as required, to make appointments and write patient assessments, notes and reports. One of the patients in this aged care facility was Patient A. Patient A had been admitted with a history of vascular dementia, frequent agitation, previous cerebral infarcts and a possible urinary tract infection.

On the day in question Patient A's daughter rang to speak to her mother whilst her mother was in the foyer waiting to be taken to see a podiatrist. The patient's daughter found that her mother was making grunting sounds and did not seem to be able to hear her, so she phoned the facility receptionist and asked for a doctor to see her mother. The receptionist asked the RN on duty to check the patient. Point 40 in the court transcript documented that "Patient A's carer (daughter) rings on a regular basis and always thinks her mother's having a stroke." After being contacted by the receptionist, the RN checked Patient A and found her having her afternoon tea. Michelle, the registered nurse, took Patient A's observations and found them "all ok" (point 22 court transcripts). As the daughter was worried, the RN tried, without success, to ring the appropriate medical practitioner to see this patient. Michelle also spoke to the clinical care coordinator who told her that Patient A had been seen by a medical practitioner who suspected Patient A had a urinary tract infection. A midstream specimen of urine had been collected from Patient A and sent to pathology to confirm the diagnosis. Pathology results from this urine sample were expected that afternoon or early the next morning.

Patient A's daughter came into the facility and found her mother "unable to speak, hear or focus" (point 17 court transcript). The daughter then spoke to Michelle, the registered nurse, who said that she had just taken Patient A's observations and needed to write her patient care nursing reports, patient assessments and resident classification documents. It was 5 minutes before Michelle was due to finish her nursing shift and she was expected to complete all of this documentation for the 46 patients before completing her work shift. Michelle did pause in finishing her paper work and take Patient A's temperature, which was satisfactory. She asked the enrolled nurse, whose job it was to take patient observations, to take the rest of the observations and report these to her as she needed to complete the paperwork before the end of her work shift. The daughter was not satisfied so she rang the medical practitioner's surgery but found that the medical practitioner was unable to talk to her as he was too busy seeing other patients. The daughter contacted the director of nursing who did come. Michelle then stopped writing her patient care notes and commenced taking a full set of neurological observations. She was taking the patient's blood pressure when Patient A had a fit. The daughter had called an ambulance. The ambulance took Patient A to Wyong Hospital where investigations, including a CT scan, found no evidence of a current cerebrovascular accident (CVA; stroke). Patient A was treated with intravenous antibiotics for her urinary tract infection and discharged from hospital 8 days later.

Patient A's daughter made a formal complaint to the Nursing and Midwifery Tribunal of New South Wales about Michelle Bottle's work as a registered nurse that day. This Tribunal found Michelle guilty of not providing an appropriate standard of nursing care and ordered her to complete, within 2 years, a Graduate Certificate in Aged Care Nursing at the New South Wales College of Nursing. There was *naming, blaming and retraining* of this nurse similar to the outcome of the equipment use error action previously described. What is interesting about this case is that in court, the peer reviewer expressed concerns "that RN's in aged care facilities are being given excessive workloads that are unsafe and unacceptable" (point 29 in court transcript) and that the employer had no orders made about their duty of care to provide safe work processes for registered nurses. Instead, it was documented that the registered nurse's professional responsibilities were not limited by having an unreasonable workload (point 62 in court transcript).

In the case *HCCC v Gallardo*, 2011, a patient's wife made a complaint about the work of registered nurse Marjorie Gallardo in relation to the nursing care given to her 70-year-old husband (called 'Patient B') on the night that he returned to the Extended Day Only (EDO) Unit following a laparoscopic cholecystectomy at Liverpool Hospital. Patient B had been transferred at 1130 hours to the EDO Unit from the operating room following his surgery. At 1700 hours Patient B was assessed by a resident medical officer (RMO) following the patient having abdominal distension and complaining of abdominal pain. The RMO (who was the patient's treating surgeon) ordered increased analgesia (15 mg of morphine) to control this patient's abdominal pain and documented that the patent's distended abdomen was due to previous surgery for an incision hernia.

On the night duty shift RN Marjorie Gallardo was the only member of the staff rostered to work in the EDO unit to care for eight post-operative patients at the commencement of her night shift. She admitted a further two post-operative patients to the EDO unit during this shift so had 10 postoperative patients to care for overnight. On this particular shift Marjorie did not have time for her usual practice of documenting all patients' observations as she was very busy dealing with one patient who was vomiting and another patient who was "distressed (crying and wailing) at being placed in a room with men, due to her religious beliefs, an issue that had not been resolved by hospital management during the earlier shift" (point 56 court transcript).

Marjorie was unable to complete all of the work required to care for all her post-operative patients so she rang the assistant director of nursing and contacted RN Roberts, who was in-charge of this unit at the time and who was working in the recovery ward, to ask for nursing care assistance. RN Tsin was sent from the recovery ward to take the post-operative patients' midnight observations. RN Tsin took Patient B's temperature, which was 38.7°C, pulse (126), respirations (16), and blood pressure (110/70 approx.). RN Tsin reported to Marjorie that patient B had an elevated temperature, but did not report that this patient also had tachycardia and a decrease in blood pressure. RN Tsin also did not report to RN Roberts, who was in-charge of this unit, Patient B's observations. A problem that RN Tsin had was that none of the patients' notes were kept at the end of their bed so she recorded all patients' observations on a piece of paper. In the EDO Unit at night there was insufficient lighting for nurses to be able to see to record patient's observations so nurses working night duty on this ward had to go to another room to record patients' observations and any other information in the patients' notes. Both RN Marjorie Gallardo and RN Tsin did use another room to write the patients' clinical notes during the night shift.

RN Marjorie Gallardo responded to RN Tsin's information by removing Patient B's blanket to reduce his temperature as she considered that the room was stuffy. She then tried unsuccessfully to page the night registrar to let him know Patient B's temperature, but this medical practitioner did not answer his page. RN Marjorie Gallardo then gave Patient B the drug paracetamol, which was due at this time, and which did lower the patient's temperature which she checked again at 1 am. RN Gallardo again paged the night registrar after taking Patient B's temperature but this medical practitioner was still not answering his page. Marjorie rechecked Patient B's temperature at 3 am and 4 am at which times this patient did not have an elevated temperature. Somewhere between 3 am and 3:30 am RN Marjorie Gallardo was able to contact the night registrar and update him on Patient B's medical condition. The night register did not come to the EDO unit to check Patient B but gave medical orders to continue to monitor this patient's temperature and phone the registrar if Patient B's temperature became elevated again. Throughout the night RN Gallardo kept a record of the patients' observations on her nursing handover sheet as patients' notes were not kept at the patients' bedside, but in another room. As she was very busy caring for 10 post-operative patients she failed to record all of the observations on her handover sheet in Patient B's nursing notes. Her nursing handover sheet was tendered as evidence in court that she had made these observations that were recorded next to Patient B's name and unit number on this sheet.

At 4:30 am RN Marjorie Gallardo took a full set of observations on Patient B and conducted an abdominal assessment. This patient had no abdominal distension at this time. RN Gallardo stated that Patient B had been asymptomatic between 12 midnight and 6 am. At 6 am RN Tsin came back to the EDO Unit to take the patients' 6 am observations. When she took Patient B's observations his blood pressure was 80/46. She reported this to RN Marjorie Gallardo who took charge of the patient's care and called the medical emergency team. Patient B was conscious. He was taken back to the operating theatre where a perforated bowel was identified. When Patient B was taken to the operating room his nursing notes went with him. This contributed to RN Gallardo not transcribing all of this patient's observations from her handover sheet into Patient B's notes as they were not available in the unit for the nurse to write this information in the patient's notes.

In court the Health Service change manager stated, "I'm once again very sympathetic of the situation with the workload on that evening, but at the end of the day the final capability is documenting it. If we have not documented it, we don't really have any evidence that the vital signs were actually carried out" (point 32 court transcript).

In court RN Marjorie Gallardo stated that "when the EDO unit had opened in early May 2007, simple ambulant self-care patients were admitted. Over time, this changed to a heavier workload" (point 66, court transcript). Mitigating factors for her actions raised by Marjory were that she was provided with minimal assistance by RN Tsin, when she asked both the assistant director of nursing and RN Roberts in the Recovery Ward, for nursing staff to help with patient care; "the heavy patient load of post-surgical patients, the Respondent's choice to provide 'hands-on' nursing rather than maintain records as she was not physically able to perform all of the tasks required of her" (point 73 court transcript).

In court RN Marjorie Gallardo had references from Clinical Nurse Specialist Gonzales and six other registered nurses who had all worked with her at this health service for a number of years. These seven nurses stated that Marjorie Gallardo was "dedicated to her work" "whilst providing a high standard of nursing care," "has demonstrated the ability to share knowledge/skills and preceptor less experienced nurses," "is always ready to help anyone in need, which is seen continuously in the caring of her perioperative patients and in the teamwork she provides to other perioperative staff members," is "thorough and precise in her work," is focused on "achieving the best outcome for both her patients and colleagues, whilst delivering a high standard of care," is "able to seek clarification when scenarios or situations are unclear," "is a good listener and a good communicator," has good professional relationships with all staff that she works with, is "a mentor to other nurses and highly regarded by all her colleagues," is "quick thinking," is "a caring, responsible, enthusiastic and kind person," has a "thorough knowledge, good judgement and very good skills as a registered nurse," is "a very committed and dedicated nurse who works with a great degree of skill and good judgement," "is hard working and values the ethics associated with honesty and integrity," "exercises responsibility in caring for people and carrying out her duty," "is an asset to her workplace, her family and indeed the whole community" and that Marjorie Gallardo "exhibits a professional standard that others strive to emulate." RN Defilippi summed up Marjorie Gallardo nursing skills with the following statement: "Marjorie displays a caring professional manner when dealing with her patients and is proactive in meeting their clinical and personal needs. The work of a registered nurse is both stressful and demanding. For me it is reassuring to have the support of my colleagues with the professional ability and understanding that Marjorie displays" (points 37–52 court transcript).

Despite the references provided by her co-workers about the high level of nursing skills that RN Marjorie Gallardo had, following this court case the Tribunal made the following orders.

1. RN Gallardo was formally reprimanded.
2. She had to complete all components of two courses of study offered by the College of Nursing. (a) The deteriorating patient: Clinical decision making. (b) Advanced assessment skills. These two courses were required to be completed within 18 months.

3. RN Gallardo was not allowed to work in charge of a nursing shift or as a team leader until she had passed the above two courses.

The results of this court case were *naming, blaming and retraining* the nurse involved. Reason (2000, p. 769) wrote "It is often the best people who make the worst mistakes." Reason (2000) recommended that the system of work should be made as safe as possible by the employer, rather than just blaming the person who made a mistake through an *active failure* when there were *latent conditions* [such as "time pressure, understaffing" (p. 769)] controlled by management that affected safe work practices.

Similar to the previously described case (HCCC v Bottle, 2011) and the patient fatality due to incorrect use of equipment (Jansz, 2011), there was no consideration of the fact that the employer had a responsibility to provide enough nursing staff for the nurse to be able to perform the work safely. In the two law cases the cause of a patient not having the level of care required could have been solve with the health care facility providing another registered nurse to work during the work shift. After the complaint of Patient B's relative, the Liverpool Hospital did change the staffing level in the EDO Unit from one to two RNs for each shift (point 89, court transcript). Looking at the court-provided references to RN Gallardo's nursing care skills, she was an outstandingly good nurse, but it was evident that on this shift, there was too much work for one nurse to complete.

All employers have a responsibility to have a safe workplace. Inadequate lighting did not make the workplace safe. Not having each patient's nursing notes next to their bed made it difficult for nurses who took patient observations to record them in each patient's nursing notes immediately and made it difficult for a nurse who had not taken patient observations to check these observations. As for the previous case a medical practitioner had checked the patient before the nurse's shift commenced but did not examine the patient during the shift. In both cases the registered nurse had difficulty in getting the medical practitioner to respond to phone calls to provide patient care advice. There seemed to be inadequate medical staff available for patient care, and this impacted on the nurse's actions and the patients' relatives' satisfaction with the nursing care provided. In both cases the nurse was required to complete more studies to become a competent practitioner without the management causes (employer's responsibility) being addressed by the tribunal recommendations.

The agency nurse is frequently allocated to work many different places in many different nursing specialities. The normal nursing shift of work is 8 hours. To save money many hospitals expect agency nurses to do the same amount of work in 6 hours as a host employer nurses complete in an 8-hour shift. Agency contract nurses are sometimes given the patients requiring the most care and who are the most difficult to care for, so that the regular nurses have a break from these patients. In some workplaces nurses are so busy with their own work that they do not help the agency nurse when help is asked for to move heavy patients, or use new equipment, and do not provide information on the correct procedures specific to the health service (such as admission paperwork, post-operative and discharge procedures) to be used for patient care in their health service. The agency nurse is then reported for not working fast enough if all patient care is not completed at the end of the work shift and if all work duties are not completed on time because they are not willing to do what is unsafe. If the agency nurse makes a mistake in this unfamiliar environment, then it is considered the agency nurse's fault. This is despite the fact that the host health service that the agency nurse is working at has a duty of care to provide safe work processes.

The following case study comes from Bakker (2012, p. 100).

> Kylie was a newly registered nurse on a surgical ward in a suburban hospital and reported an occasion when she was rostered from 10 am–6 pm. She was originally given five post-operative patients for whom she needed to do regular observations. A staff member went off duty at 12.30 pm leaving Kylie with two other patients. The other two patients were delegated to a casual nurse, but it was found out later that this nurse was not due to start work for another hour. When this staff member arrived she reported the shortfall of care and Kylie was reprimanded in the middle of the ward by her coordinator for not taking care of the two patients. "This made me feel guilty and horrified at the neglect of care that occurred apparently on my behalf. . . . I was also devastated that an incident report had been filed citing me as responsible." As well as this, she had not had a lunch break. Kylie felt she had been "abused" by her senior staff and, after a similar episode several weeks later, she asked to be transferred to another ward.

This situation can occur with agency nurses. Many times agency nurses do not have a meal break (or even a toilet break) as they are too busy caring for their patients and worry that they cannot get everything completed in the allocated time, even when they work unpaid overtime to be able to complete all of the nursing care required for their shift and to complete all of the documentation of care given. The story that Kylie tells above is what can happen when there is a gap between when the agency nurse completes her shift of work and when the next shift of work nurse commences her work, which is often 2 or more hours later. The host employer has a duty of care to provide enough staff for safe patient care.

An example of good work organisation is at Bethesda Hospital where, on the palliative care ward, nurses work as a team of three. This work organisation has many benefits. It means that when there are sick patients, as in the case of *HCCC v Gallardo*, 2011, one nurse can tend to the patients that require the extra care whilst the other nurses continue on with giving medications, taking nursing care observations and completing the required patient care work on time. It allows enough nurses to care for the patients so that both the agency nurse and the host employer nurses can all go, one at a time, to have meal and toilet breaks and that there is a nurse who know the patients to care for them when the agency nurse finishes her shift if this person has a shortened shift of work. Having this work organisation helps to provide adequate supervision for the agency nurse and to provide the agency nurse with the required information to work safely and effectively. This work organisation provides the agency nurse with someone to ask if the agency nurse needs to know where equipment is kept, how to use equipment safely, how to care for individual patients (particularly if the patient can be aggressive, manipulative or very heavy to move), the way that dressings and other patient care treatment is to be performed at this health service and the correct paperwork to complete for patients' admission, health care and discharge. If the agency nurse works alone, then it can often take 30 minutes or more for another nurse to be available to help to move a heavy patient using a lifting hoist (which is unsafe for one person to use on their own), to check drugs of addiction with, or for assistance with nursing care which requires two or more people. This decreases productivity as well as nurse and patient safety.

Agency nurses need enough information to be able to work safely. The case of *HCCC v Bottle*, 2011 was in relation to a nurse who worked in a nursing home. Working in nursing homes can be very difficult for agency nurses as none of the patients have any form of identification on their person. In a hospital all patients wear a wristband with their name, date of birth, and patient unit number so that they can be safely identified. An agency nurse who is new to caring for a patient can check the patient's name, birth date, and patient unit number with each person's medical record to ensure that the correct drug and treatment is given to the correct patient. In a nursing home, some patients (residents) may be confused, or not able to provide accurate information about themselves. Some nursing homes have a photo of each patient on the patient's medication chart, but the photo may be out of date and the resident looks nothing like their photo. To meet their duty of care in providing enough information for the agency nurse to work safely it would be an advantage to give each nursing home resident a bracelet or watch band with their name, date of birth and a patient care number to wear so that the agency nurse can identify each person to give them the correct medication and care. Under the present system, for a nurse new to the nursing home, the nurse has to ask for a permanent staff member to verify who each resident patient is to ensure that correct medication and correct care is provided to each person. This decreases productivity.

An example of a good system of work was identified at Anchorage Aged Care. At this health care facility, during their orientation, agency nurses are introduced to the patients that they will care for during their work shift so that they can recognise them and know the location of each resident's room. Typed patient handover sheets are given to each agency nurse. The sheet has each patient's name and relevant present-care requirement medical information, which is handed over to the agency nurse orally (as well as in written form) by the nurse who cared for each resident during the previous work shift. This enables the agency nurse to ask questions about resident care requirements if more information is needed to be able to work safely. As an error-reducing (and time-saving) strategy, all patients' medications are in a medication packet with the correct medication to be given for each time. Patients' medications are put into these pockets once a week by a pharmacist. To reduce documentation, nursing notes are only written by the agency nurse for patients who had any change in their medical condition during the shift worked. This allows more time for hands-on nursing care. The nurse manager at this facility is very supportive in caring for all of the staff working at Anchorage Aged Care and extends this support to agency nurses when they requested help for situations that they were unfamiliar with. Agency nurses who worked well caring for the residents are asked by the nurse manager to come back to work at the facility the next day and when they are available for work. For the health service this provides continuity of patient care. For agency nurses this meant that they are more familiar with the residents that are to be cared for and with the work processes to be completed, that they have more of the information required to be able to work safely and that they are able to provide a high standard of nursing care.

Evaluation and continuous improvement

A way for health care services to evaluate what they do well, and where there are opportunities to improve patient and staff care and safety, would be for health services to ask agency nurses for feedback about their work shift and to involve agency nurses in their quality and safety activities. Agency nurses work at many health services, so see evidence of best practices in work organisation, staff and patient care, and safety practices, which they could share with other health services so that evidence-based best practices can be implemented as appropriate. They see the organisation and the work practices as an outsider so can identify hazards and risk control measures that staff that work at the health service on a daily basis may be blind to. In a research study titled 'An Analysis of Quality

Practices and Business Outcomes in Western Australian Hospitals', Mussett (2001) noted that the hospitals with the highest standards of safety and patient and staff care had a mission and a culture of caring for everyone who came to the work premises. This included caring for agency nurses. The findings of this research have been found to be applicable to all industries and are in line with the International Labour Organisation (ILO) Convention 155, which promotes a general duty of care for everyone who comes onto the work premises and/or who can be affected by workplace activities, products or services. The factors in this model are included in Figure 5.2. In this figure, Level 4 Quality Activities mean

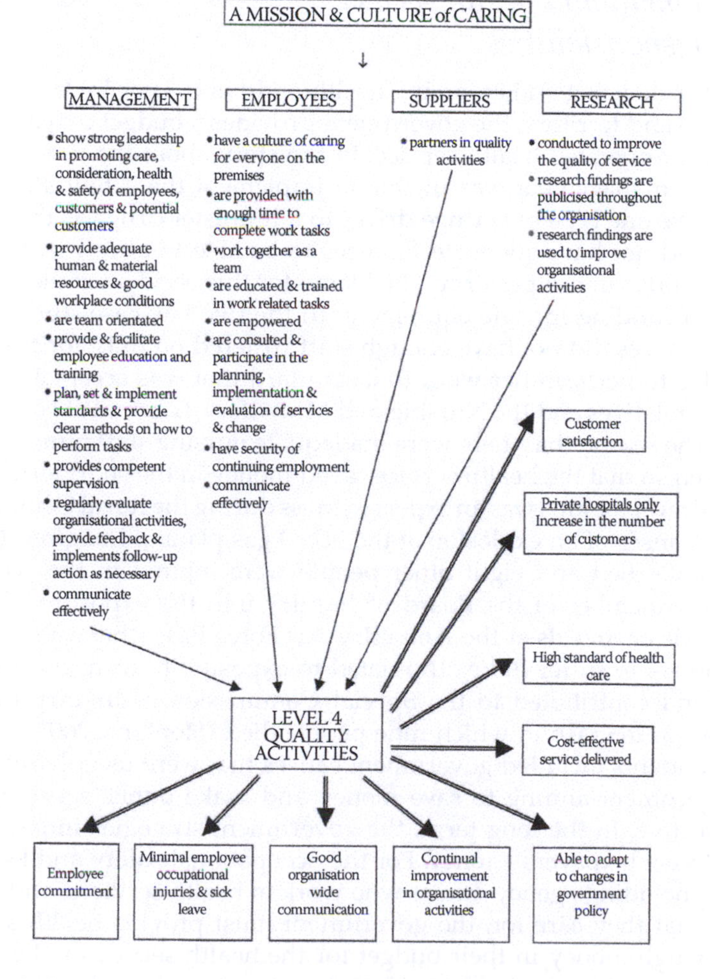

Figure 5.2 Quality care model. (From Mussett, J. 2001.)

'quality anticipation.' Level 4 quality activities surprise and delight customers, exceed customers' expectations and provide new innovations by identifying customers' unconscious need and expectations. At this level of quality activities, employees are empowered, asked to be innovators and work towards a common company vision.

In this model, the nursing agency is the supplier of human resources to the host health service. If management and employees are performing all of the actions in this Quality Care Model, then an outcome is a high standard of health and safety practices for the host's own employees, for agency nurses and for everyone who comes onto the business premises. The host employer has then met its duty of care.

Government and health service board responsibilities

As well as devising and enforcing health and safety laws in Australia for each state and territory, the government provides a budget for each state- or territory-operated health service. In the story about the use of equipment, top management was unable to implement the safety suggestion to just have one type of syringe driver to administer drugs as this health service did not have adequate finances to be able to afford making the nurses' work safer to perform. The blame for this accident could go back to lack of funding for safe equipment. In the two law case studies, both health services did not have enough staff rostered on duty for the nurse to be able to perform her work to a standard that was acceptable to the patients' relatives and the Nursing and Midwifery Tribunal of New South Wales. The reason that there were inadequate nursing staff numbers may have been so that the health service saved money in the short term.

Andrew Hopkins was an expert witness during the Royal Commission into the causes of an explosion at the ESSO gas plant at Longford (where two people died and eight other people were injured in the accident); an expert member of the Board of Inquiry into the exposure of workers to toxic chemicals at the Amberley Air Force Base (that caused about 600 workers to suffer ill health related to exposure to toxic chemicals at work); and contributed to the Special Commission of Inquiry into the waterfall train crash in which nine people died (Hopkins, 2005). Each of these accidents included government causes that were usually related to the government aiming to save money and make public services more cost-effective. In the long term, the government strategies implemented did not save taxpayers' money. For the occupational safety and health of people, including agency nurses who work in health services, and for the people that they care for, the government must provide health services with enough money in their budget for the health services to be able to afford to make the workplace and work processes safe and for the health services to be able to afford to purchase the products, equipment and services of people that are required for workplace safety and health.

Whilst government health services are influenced by government policies, procedures and the finances provided by the government, finances and often the culture of the organisation of private health services are usually controlled by a board of management. Mussett (2001) identified that a private health service, with a very high standard of health and safety care for all of its customers and staff, including agency nurses, had a management structure in which the Health Service Board of Management was at the bottom of the organisational structure. Above the Board of Management was the Chief Executive Officer (CEO) who was supported in his work by the Board of Management Members. Above the CEO were the Director of Nursing and the Hospital Administrator who were supported in their work by the CEO. Above these two people were the other hospital management employees who were supported by the Hospital Administrator and Director of Nursing as appropriate. Above these management staff were the nurses and other people who provided the patients' hands-on care. These workers were supported by their line manager, which, in the case of the agency nurses, was usually the nurse co-ordinating the ward nursing staff on the shift that they worked at the health service. Support included having enough of the right equipment, products and staff to provide a high standard of patient care. Agency nurses who worked at this health service were given an appropriate orientation of where equipment, policy and procedures that they will be likely to use was located and were provided with ongoing information support throughout the agency nurses' shift of work. At the top of the organisational chart were the health service customers who were provided with a very high standard of health care.

The room with the best view in this health service was the staff room as it was considered by the Board Members that if staff were well cared for they, in turn, would be able to provide a high standard of health care to patients. All staff members, including agency nurses, were provided with food and drink at their tea and meal breaks. This hospital looked at caring for patients (customers) and staff members' physical, mental, social and spiritual needs. Agency nurses requested to work at this hospital when there was an opportunity. At the end of the shift patients in this health service frequently said "thank you" to the nurse caring for them during the shift and asked if this nurse would be back to care for them the next day.

Do patients and residents have health and safety responsibilities?

Patients and nursing home residents can cause agency nurses injury either intentionally or unintentionally. Intentional injuries occur to nurses when patients or residents are aggressive. This is one of the reasons that all agency nurses must have annual lectures (or pass an annual online challenge test) on how to deal with aggressive patients, residents, and other people who visit the health service and may be aggressive towards nurses. These lectures on their own are not effective in dealing with all aggression from patients, their relatives and health care visitors, so many health services have a 'Code Black' where an agency nurse, or other staff member, can call for help from trained personnel in the health service to assist in dealing with physical violence directed at them. Unintentional injuries can also occur such as when an agency nurse is showering a patient and the patient falls over, causing undue stress on the nurse's body.

Patients, residents, their relatives and visitors to the hospital have a responsibility not to interfere with, or misuse, anything in the health care facility that is provided for occupational safety and health (such as propping open a fire door). They are "not to create false health and safety fears with the aim of disrupting the workplace" (New South Wales Nurses' Association, 2004, p. 15). Patients, residents, their relatives and visitors to the hospital have a common law responsibility not to inflict harm on other people at the workplace by any acts or omissions.

Nursing agency duty of care

As well as the host employer having a duty of care to the agency nurse, the agency that the nurse works for has a duty of care. The nursing agency has the responsibility to ensure that all nurses that it employs to work are registered with the Australian Health Practitioner Regulation Agency for the employment positions that they will work in. This can easily be checked on the Internet as all nurses' registration is available for the public to view. They need to check that the nurse has completed all of the criminal screening and educational competencies required by each health service and that each nurse completes at least 20 hours of nursing education each year to keep up to date with current nursing knowledge. All of these are required to ensure that the nurse will care for patients safely.

The agency that the contract nurse works for does not have control of the host employer's physical workspace, equipment required for health care use, work organisation, the policies, procedures and guidelines. However, the agency does have a legal requirement to check that the place that the contract nurse is employed for the shift at is a safe workplace with safe work processes. For this reason the nursing agency manager, or a representative with an expert knowledge of workplace safety, should inspect the work premises, equipment and products used at each facility that agency contract nurses will be employed to work at. Work processes are best understood by the nurses who will be doing the work. As part of their tertiary education nurses are taught to work safely. The best way to ensure that the work practices are safe is for the agency to encourage the agency nurses to provide work shift experience feedback to give praise for what was well done in relation to the work organisation and work processes, as well as to identify opportunities for improvements if these are required at each health service that they work. This feedback then needs to be provided to the facilities manager that the contract nurse worked for. An agency may even consider providing an annual award to the health service that its staff voted to be the safest and healthiest place to work.

Safe Work Australia (2012, p. 3–4) records that the duties that a labour-hire agency needs to complete to ensure their employees' health and safety before they commence work with a host employer are as follows.

- Review the host employer's safety record to identify if the workplace is safe to send agency nurses to work there.
- Gather information about the work and the workplace, including the work environment, organisational arrangements, health and safety risks associated with the work and any skills and knowledge the worker will require to safely undertake the work (p. 3).
- Verify that the host employer will provide site-specific and task-specific induction, training and personal protective equipment, if this is required, to agency nurses.
- Assess the workplace for any risks to health and safety. Where risks are identified, consult with the host to ensure they are eliminated, or if that is not reasonably practicable, minimised (p. 3).
- Ensure that workers have the necessary qualifications, licences, skills and training to safely carry out the work (p. 3).
- Consult with the host employer and workers to ensure that the agency and the agency nurses understand, and are confident in the understanding, of the work safety and health policies, procedures and practices of the host employer.
- Ensure that agency nurses have the means to identify and take action in an unsafe situation at the host employer's workplace, such as notifying the host employer manager, or other staff, or stopping work until the situation is safe.
- Establish communication methods that the agency nurses can use to contact their agency if they consider that there is a risk to their safety or health.
- Ensure that agency nurses are able to raise issues with their agency if they are not satisfied with the host employer's response to their concerns.
- Take effective action if an agency nurse or host employer identifies risks, or raises concerns, about workplace health and safety matters.
- Provide agency nurses with a general workplace health and safety induction and training that cover any risks identified at the host

employers' workplaces and the consultation methods that are established between agency nurses and the host employers.
- Encourage agency nurses to maintain contact with their agency and provide feedback about health and safety issues or concerns in the host employers' workplaces.
- Ensure that there is a documented system in place for the management of work safety and health.

These recommendations should be used, where practical, to ensure contract employees' workplace safety and health are not compromised.

Hours of work

Fatigue due to hours of work is a problem for many hospital-based nurses as they are often required to work an early shift after a late shift which, with the time needed for travelling, eating and meeting hygiene needs included, does not always allow the nurse to have sufficient sleep between shifts. Hospital-based nurses may also work physically hard for up to 10 days in a row with no days off work. Hospital-based nurses may be required to work set blocks of time on night duty and, due to not being accustomed to sleeping during the day, may not sleep well and be tired when working at night. Host employer nurses may work frequent overtime if staff numbers are low. When a nurse is fatigued, errors often occur (Reason, 1990).

A reason that some nurses choose to work as a contract agency nurse is because they can choose the hours that they work. With a choice of working hours agency nurses can choose to work shifts where they do obtain enough sleep and at times that fit in with what else they are doing in their life. For example, caring for school-aged children, studying at a tertiary institution, or working in another employment position. Having a choice of work hours allows agency nurses to work at the times when they are alert, and this enhances their ability to work safely. The responsibility for fatigue management of contract nurses lies with both the nurse (who chooses the shift times to work) and the agency who allocates the work to the nurse.

Demonstrating care for staff

As part of their health and safety responsibilities the nursing agency needs to demonstrate care for their staff. In 2010 CPE Group won the award of "Employer of the Year", based on employee feedback. At this agency, nurses have a choice of where they would like to work, depending on the work available and the nursing skills that they have. This promotes good mental health as these contract nurses are able to work in health services that they enjoy. It allows continuity of knowing more about the workplace, work processes and the people they will work with, for nurses who choose to go to places they are familiar with. A research study titled "An Evaluation of Work-Related Stress in Nurses" (Mussett, 1991) identified that nurses who worked in more than one area, such as pool and agency nurses, had the least amount of stress. These nurses were the least likely to view their job as affecting their health or to feel restricted by over-supervision, and the most likely to find their work interesting and to look forward to coming to work. When nurses working for an agency are well cared for by their employer, the findings are similar (Jansz and Mills, 2008).

CPE Healthcare Group provides social activities as well as an online community through Facebook for its contract nursing staff so that these nurses can get to know their management staff and each other. The agency provides small gifts (such as a diary to use for work and a pen with the company name on it) for each contract nurse for Christmas. In its newsletter the CPE Healthcare Group publishes a congratulatory article and a photo of the employee of the quarter award winner every three months. This is the nurse who has received the most praise from the staff that they have worked with, helps share this success, and encourages the celebration of excellence in nursing performance. As well as being effective in promoting social health, this helps the nursing agency to retain good nursing staff to send to host employers, has helped this agency and has assisted it to establish an excellent reputation.

Agency nurses' responsibilities for health and safety

An agency nurse needs to be an experienced nurse who is competent in working in many different health services, caring for patients with many different medical and surgical health care requirements. A key attribute of agency nurses is feeling comfortable asking for information, being a good communicator, and being adaptable to the situations in which they find themselves. The agency nurse needs to be able to identify hazards, assess the risk of each hazard causing harm, control the risk of identified hazards causing harm and to report any hazards to the person co-ordinating the area that they are working if the agency nurse cannot control the risk of harm occurring.

As well as having a duty of care to ensure their own health and safety at work, the agency nurse has a duty of care to avoid adversely affecting the health or safety of any other person through any act or omission of what they do at the workplace. For this reason agency nurses, for any situation that they are unsure about, need to have a competent person to ask questions to, ensuring that they work safely, within the limits of their competencies. If there is no host employee to ask, then the agency nurse needs to ring her agency, explain the situation, and ask for advice.

Agency nurses must wear any host–employer-supplied personal protective equipment that is required for nursing care. This is particularly important in infection control situations. The agency nurse should follow instructions given by the host employer's employees (who are familiar with the workplace, work processes and people to be cared for), when it is safe to do so. If the agency nurse does become injured or ill at work this nurse should then report the injury or ill health to both the host employer and the agency.

Agency nurses have the choice of accepting, or not accepting, work at any health service when work is available. If they feel that the situation in any health service is not safe, they can say no to accepting work at this health service. Conversely, health services have the right to state which agency nurses they will have working at their health service and only accept nurses that they want to work there.

Survey information on agency nursing

In 2003 the Australian Nursing Federation (Western Australian Branch) conducted a survey related to agency nursing. More than 500 nurses replied. The exact number of respondents was not provided, and results were reported as percentage (%) only. There has not been a more recent survey of agency nurses' work published, but the results of the findings of this research still reflect the situation today in Australia. Of the nurses who replied who had worked as an agency nurse during the last 12 months, 25% stated that agency nursing was their only source of employment whilst 72% documented that they worked agency nursing in addition to having another employment position. Another group comprising 51% recorded that they worked agency nursing shifts in the same hospital as they normally worked. To be able to work as an agency nurse, 23% wrote that they had decreased their permanent employment hours of work in a hospital.

When asked why they chose to work as an agency nurse, the top 3 answers were as follows.

Better pay (69%).
Choice of shifts (68%). Time off when you want it and the choice of shifts to fit in with your family, childcare, or your lifestyle.
Workload management (25%) (Olson, 2003, p. 10).

The majority of nurses working agency indicated that they enjoyed not being part of ward politics.

As well as surveying agency nurses the opinions of nurses who worked with agency nurses were also recorded. The main themes that these nurses wrote about concerning the negative impact of having agency nurses were as follows.

> Rostered 8-hour shifts being replaced with 6- or 7-hour agency shifts—with permanent staff left to carry the missing hour or two of nursing.
>
> Lack of continuity of care.
>
> Permanent staff required to spend more time orientating new staff, often on every shift.
>
> Flexibility and higher pay associated with agency nursing generates discontent among regular staff (Olson, 2003, p. 11). (The reason that agency nurses have better pay in the short term is that this is to compensate for not being eligible for sick leave, annual leave, long-service leave and other leave payments.)

One nurse summed up the situation as "It's better to have an agency nurse than no nurse at all but it takes longer to get the work done as you have to explain and show them where things are" (p. 12).

The positive aspects of employing agency nurses for patients' and/or residents' care were documented by nurses who worked in private and public hospitals and who worked in nursing homes as follows.

> They maintain basis numbers and if they are regulars (or actual staff from the ward doing an extra shift through the agency), then the perception is it's just like having regular staff.
>
> They allow hospitals to respond quickly to unexpected demands, particularly in the critical areas.
>
> They bring additional skills, knowledge and new ideas to ward/unit/nursing home. (Olson, 2003, p. 11)

As a summary of the positive aspects of employing agency nurses, the following comments were recorded by hospital and nursing home permanent staff:

> "Most agency staff do excellent work and fill shortages to help maintain care to patients. They reduce the stress for permanent staff."
>
> "We had two agency nurses working in our ward for a month each. I have only praise for their expertise and the speed in which they settled in."
>
> "When I have had the good fortune to work with agency staff they are not weighed down by the 'crap' of the permanent staff. They are much happier."
>
> "These people are very useful and welcome. You learn things from them and it reminds you that your way isn't the only way" (p. 12).

Summary

Working as an agency nurse can be rewarding as the nurse meets many interesting people, both staff and patients, is constantly learning and adapting to new situations and is often thanked by the host employers' staff and patients for their work. In turn, if the agency nurse has worked with supportive staff, this nurse usually thanks the people he or she has worked with for their work. The management of health and safety for contract nurses is the responsibility of the host employer, the host employer's employees who work with the contract nurse, the agency that the contract nurse works for and the contract nurses. Good contractor safety management is important for the nursing agency that employs nurses, for the host employer, for the customers that the contract nurse cares for, and for the agency nurses themselves.

References

An Analysis of Quality Practices and Business Outcomes in Western Australian Hospitals (Doctoral dissertation). Edith Cowan University, Joondalup, WA, p. 238).

Ashley, J. (1976). *Hospitals, Paternalism and the Role of the Nurse*. Columbia University, New York. Teachers' College Press. Retrieved from: http://www.nhmrc.gov/files_nhmrc/publications.

Australian Government. (2009). Health Practitioner Regulation National Law Bill Explanatory Notes. Available from http://www.ahpra.gov.au/Legislation-and-Publications/Legislation.aspx.

Bakker, S. (2012). Covert Violence in Nursing—A Western Australian Experience. (Doctoral dissertation). Edith Cowan University, Joondalup, WA.

Commission of the European Communities. (2002). *Adapting to Changes in Work and Society: A New Community Strategy on Health and Safety at Work 2002–2006*. Brussels, Belgium: Author.

Department of the Prime Minister and Cabinet. (2004). *Women in Australia 2004*. Canberra, ACT: Australian Government.

Evaluating the Effect of Technology into the National Airspace System. Retrieved from http://rutgersscholar.rutgers.edu/volume05/luxhoj-kauffeld/luxhoj-kauffeld.htm., p. 1).

HCCC v Bottle [2011] NSWNMT 3 (2011, March). Retrieved from http://corrigan.austlii.edu.au/au/cases/nsw/NSWNMT/2011/3.html.

HCCC v Gallardo [2011] NSWNMT 8 (2011, April). Retrieved from http://corrigan.austlii.edu.au/au/cases/nsw/NSWNMT/2011/8.html.

Hopkins, A. (2000). *Lessons from Longford: The ESSO Gas Plant Explosion*. Sydney, NSW: CCH Australia Ltd.

Hopkins, A. (2002). Lessons from Longford: The trial. *The Journal of Occupational Health and Safety. Australia and New Zealand, 18*(6), 5–67.

Hopkins, A. (2005). *Safety, Culture and Risk: The Organisational Causes of Disaster*. Sydney, NSW: CCH Australia Ltd.

International Labour Organisation. (1981). *Occupational Health and Safety Convention No. 155*. Geneva, Switzerland: ILO.

Jansz, J. (2011). Cognitive ergonomics. In Barrett, T. Cameron, D. and Jansz, J. (Eds.) *Safe Business: Good Business. A Practical Guide to Occupational Safety, Health and Insurance in Australasia (3nd ed.).* Guildford, WA: Vineyard Publishing Pty Ltd. pp. 107–130.

Jansz, J. and Mills, S. (2008). Occupational health and safety. In Crowther, A. (Ed.) *Nurse Managers: A Guide to Practice (2nd ed.).* Melbourne, Victoria: AUSMED Publications.

Kalisch, B. and Kalisch, P. (1975). Slaves, servants or saints? *Nursing Forum, 14*(3): 222–263.

Luxhoj, J. and Kauffeld, K. (2003). Evaluating the Effect of Technology into the National Airspace System. Retrieved from http://rutgersscholar.rutgers.edu/volume05/luxhoj-kauffeld/luxhoj-kauffeld.htm.

Mussett, J. (2001). An Analysis of Quality Practices and Business Outcomes in Western Australian Hospitals (Doctoral dissertation). Edith Cowan University, Joondalup, WA.

Mussett, J. (1991). An Evaluation of Work Related Stress in Nurses (Master dissertation). Curtin University.

New South Wales Nurses' Association. (2004). *Occupational Health and Safety Essentials for Nurses.* Retrieved from http://www.nswnurses.asn.au/multiattachments/3009/DocumentName/ohs_complete.pdf.

Nightingale, F. (1969). *Notes on Nursing: What It Is and What It Is Not* (Reprint). New York. Dover Publications.

Nursing and Midwifery Board of Australia. (2012, March). Who does what in nursing and midwifery regulation in Australia. *News for Nurses and Midwives, 1*: 2.

Olson, M. (2003, March). If conditions were good enough nurses would be permanent by choice. *The Western Nurse.* pp. 10–12.

Reason, J. (1990). *Human Error.* Cambridge, UK: Cambridge University Press.

Reason, J. (2000). Human error: models and management. *British Medical Journal. 320*: 768–770.

Reverby, S. (1987). *Ordered to Care—the Dilemma of American Nursing, 1850–1945.* Cambridge, UK: Cambridge University Press.

Safe Work Australia. (2012). Labour Hire: Duties of Persons Conducting a Business or Undertaking. Retrieved from http://www.safeworkaustralia.gov.au/AboutSafeWorkAustralia/WhatWeDo/Publications/Pages/LabourHireFactSheet.aspx.

Simkins, J. (2012a). Elizabeth Fry. Spartacus Educational. Retrieved from http://www.spartacus.schoolnet.co.uk/REfry.htm.

Simkins, J. (2012b). *Florence Nightingale.* Spartacus Educational. Retrieved from http://www.spartacus.schoolnet.co.uk/REnightingale.htm.

Tellis, W. (2003). Introduction to Case Study. The Nursing Record. 1892. London. Retrieved from http://www.tmaonline.org/whatismont.shtml.

WorkSafe Victoria. (2008). Case Studies in Labour Hire. Retrieved from http://www.worksafe.vic.gov.au/wps/wcm/connect/787776004071f3a294d9dee1fb554c40/case_study_labour_hire.pdf?MOD = AJPERES.

Zenergy Recruitment. (2012). Un-Inducted Labour-Hire Worker Awarded $750k in Damages. Humphries v Shoalhaven City Council [2012] NSWDC 216 (23 November 2012). Retrieved from http://www.zenergyrecruitment.com.au/newsletters/nov12/e.htm.

chapter six

Co-ordinating industry stakeholders to achieve safety and health excellence

Patrick B. Gilroy, AM
Chief Executive Officer, MARCSTA

Janis Jansz, PhD
Curtin University School of Public Health
Centre for Abororiginal studies
Edith Cowan University, School of Business
Curtin Health Innovation Research Institute

Contents

Introduction

This chapter has endeavoured to document the consequences of the introduction of occupational safety and health legislation in Western Australia at a time when the contracting industry began to emerge as a significant force in the industry. The outstanding feature of this combination of factors was the co-operative and positive approach taken by industry, its contractors and the regulatory authority to achieve an almost seamless transition to the adoption of the legislation. The resultant world leadership in occupational health and safety should be recognised and the method of achievement adopted by other major industry sectors who have yet to fully recognise the benefits to both their workforce and the cost reductions that can follow.

The first miners in Australia were the Aboriginal people with both men and women working as miners (*Mineral Resources,* 2007). The Chamber of Minerals and Energy of Western Australia (2012a) records that Aboriginal people have continued to work in mining, and in 2012, they made up 4.2% of the Western Australian mining industry workforce (i.e., as 3,816 Aboriginal male and female mine workers at that time).

The first commercial mining in Western Australia by people of non-Aboriginal descent commenced in 1891 with gold mining in the Murchison district, gold mining at Coolgardie in 1892, and with gold mining in Kalgoorlie in 1893 (*Australian Mining History,* 2012). The first contractors used in the Western Australian mining industry were introduced when companies were founded to mine this gold. Figure 6.1 shows the increasing use that has been made of contractors in the Western Australian mining industry from 1987 to 2008.

The total minerals and energy workforce in Western Australia in 2012 was 92,300 people employed in minerals projects and 8,300 working in petroleum projects (Chamber of Minerals and Energy of Western Australia, 2012b). Of this workforce, 23,000 were fly-in-fly-out (FIFO) workers with 79% flying in and out of Perth and the rest of the FIFO workforce coming from other parts of the state (10%) or from interstate (11%; Chamber of Minerals and Energy of Western Australia, 2012b). In 2012 workers in the mining industry were the most productive workers in Australia as is shown in Figure 6.2.

In 2012 Western Australia had 513 commercial mineral projects, 893 operating mine sites, 64 operating oil and gas fields, and had 140 exploration managers helping to identify new mining opportunities (Department

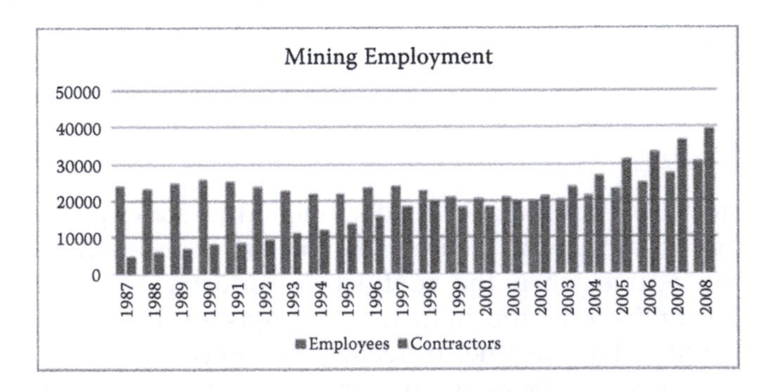

Figure 6.1 Contractor employment in the Western Australian mining industry. (From Resources Industry Training Council. (2010). *Western Australian Mining Industry: Workforce Development Plan*. Perth, WA: Author, p. 11.)

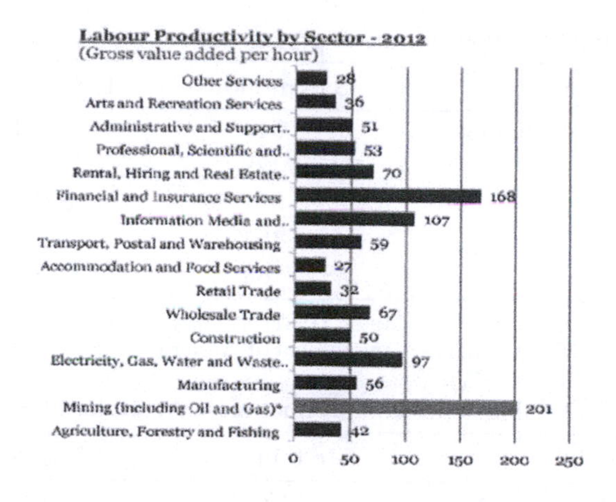

Figure 6.2 Employee productivity in Australia. (From the Chamber of Minerals and Energy of Western Australia. (2012b, November). *WA State Growth Outlook 2013*. Perth, WA: Author, p. 51.)

of Mines and Petroleum, 2012b). In the 2011 calendar year, the Western Australian mining industry was worth $107 billion (Department of Mines and Petroleum, 2012a). Major products mined in Western Australia include iron ore, gold, alumina, lead, silver, copper, vanadium, coal, oil and gas, nickel, mineral sands, clays, salt, diamonds, cobalt, chromite, manganese, talc, gypsum, pegmatite, limestone, manganese, molybdenum, tin, rare earths, phosphate, tantalum, lithium, sand, uranium, zinc and silica (Resources Industry Training Council, 2010).

This introduction has provided a concise description of the mining industry in Western Australia. In 2012 as much as 70% of work in the Western Australian mining industry was performed by contractor companies. The rest of this chapter documents how contractors have improved the health and safety management practices in the Western Australian mining industry and the factors that contributed to this success.

How occupational safety and health excellence was achieved

In 1981 the value of mineral production in Western Australia was $2.692 billion (Department of Mines, 1981) produced in the main by five sectors (Table 6.1):

Employment in the mining sector was 24,063 with 21,620 employed on the surface and 2,443 underground (Department of Mines, 1981). By 1988–89 the value of production had increased to $90 billion, which included petroleum products valued at $69.6 million (Department of Mines, 1989; Table 6.2).

Table 6.1 Value of mineral production in Australia 1981

Mineral	$ In millions	Employees
Bauxite/alumina	548	3378
Iron ore	1,129	11,847
Nickel	324	3,141
Gold	153	2,147
Coal	68	962

Source: Department of Mines. (1981). *Western Australian Annual Report 1981*. Perth, WA: Government of Western Australia, p. 157.

Table 6.2 Value of mineral production in Australia 1988–1989

Mineral	$ In billions	Employees
Bauxite/alumina	1.62	5,393
Iron ore	1.78	8,849
Nickel	0.63	3,271
Gold	2.03	9,915
Coal	0.16	1,271
Diamonds	0.36	824

Source: Department of Mines. (1989). *Western Australia 1988–1989 Yearbook*. Perth, WA: Government of Western Australia, p. 22.

In 1988–89 employment in the mining sector was 30,332 with 27,558 employed on the surface and 2,774 underground (Department of Mines, 1989). The remarkable increase in gold production, occasioned by an escalating increase in value, saw the emergence of contracting organisations usually established by senior employees of corporations who grasped the opportunity to improve their immediate and future prospects.

At the same time, the incumbent Labour Government gave notice of its intention to introduce Robens*-type occupational health and safety legislation which would require employers to co-operate and consult with employees in the implementation of health and safety policies and procedures in the workplace. Employer response to the Government's proposals was initially negative as it was concerned at the potential for increased industrial activity by unions resulting from the appointment and empowerment of health and safety representatives.

The Occupational Health and Safety Act was proclaimed in 1984 and, although the mining sector continued to operate under the Mines Regulation

* Alfred Robens (1910–1999) was an English trade unionist, Labour politician and industrialist. He spent a decade as chair of the National Coal Board and later headed a major inquiry that resulted in the 1972 Robens Report on health, safety and welfare at work, which championed the idea of self-regulation by employers. The report led to the creation of the UK Health and Safety Commission and the UKHealth and Safety Executive.

Act, there was a clear intention by government to ultimately bring mining under the general industry legislation. An alternative was to amend the Mines Regulation Act to reflect the occupational health and safety principles applying to all other industries. In anticipation of this inevitability the mining industry began to introduce some of the principles of the Robens-type legislation into its procedures and practices, and also began to work more closely with its regulatory authority in the drafting of legislation that reflected the Occupational Safety and Health Act 1984, and the considerable number of regulations necessary to reflect the legislative intent.

In 1989–90, the safety performance of the industry was acknowledged as unacceptable with a serious injury rate of 19 per 1,000 employees and a minor injuries rate of 48 per 1,000 employees. Fatalities at that time had risen to an alarming level of almost 20% per annum, occurring mainly in metalliferous underground mines (Department of Mines and Petroleum, 2013). Contractors who were carrying out much of the work were coming under close regulatory attention at this time and were frequently accused of operating with inadequate safety practices and procedures.

The controversy regarding the introduction of health and safety representatives and the empowerment of the workforce following the proclamation of the Occupational Health and Safety Act in 1984 was a significant factor in generating a positive change of attitude in the mining sector. Aware of the commitment of the government to incorporate Robens-type principles into the mining legislation, the Chamber of Mines established an Occupational Health and Safety Committee which met for the first time on August 1, 1983. The purpose of the meeting was to consider the likely effects of introducing the proposed occupational health and safety legislation into the mining industry. In the years that followed the composition of the committee was augmented to include specific expertise from all major sectors, and the committee took a positive direction that spread quickly through the regional network of similar industry committees. Expertise in the augmented central committee included audiometry, industrial hygiene, occupational medicine and safety. The committee quickly became influential in negotiating with government and, in particular, with its regulatory authority.

The Chamber's Annual Report for 1987 commented that

> through its representation on the WA Occupational Health, Safety and Welfare Commission, the Committee continued to provide positive input into decision making on health and safety in the workplace, the positive association with the Department of Mines in the preparation of appropriate amendments to the Mines Regulation Act and Regulations and its considerable input to the development of

> the new computerised Accident Reporting System (AXTAT) for the mining industry, expected to take effect from 1 January 1987. [The Chamber of Mines of Western Australia Inc., 1987, p. 11.]

A major initiative in 1990 was the organisation of the first Minesafe International conference, which was attended by representatives of 15 nations and established the Western Australian mining sector as a world leader in the discipline of occupational health and safety. Further conferences were held in Perth in 1993 and in 1996, in South Africa in 1998 and again in Perth in 2000. Papers presented at those conferences addressed every aspect of occupational health and safety in the mining industry, and presenters included contracting organisations at a number of these events. In 1997 the Chamber of Minerals and Energy produced a publication titled *A Guide to Contractor Occupational Health and Safety Management for Western Australian Mines*. This publication was developed 'to contribute to improved management of contractor occupational health and safety by mining organisations' (p. 3) and provided information about occupational health and safety considerations for contractor selection, work and contract completion.

The regulatory authority and MARCSTA*

Throughout the late 1980s and the 1990s a unique working relationship developed between industry and its regulatory authority, the Department of Mines, which became the Department of Minerals and Energy.

The appointment in 1984 by government of an experienced, highly competent mining engineer† from industry to the position of state mining engineer, at a time of industry expansion and the introduction of occupational health and safety legislation changed the nature of the previous relationship from arms-length to close co-operation in the regulation of the health and safety for the mining industry.

* 1894–1992, Department of Mines; 1992–2001, Department of Minerals and Energy; 2001–2003, Department of Mineral and Petroleum Resources; 2003–2009, Department of Industry and Resources (DOIR)—Mineral and Petroleum Resources Division; 2009–current, Department of Mines and Petroleum (DMP).

† James Milne Torlach (1938–2006; State Mining Engineer, 1984–2001) made an outstanding contribution to the improvement of safety and health in the mining industry in Western Australia, being responsible for the complete overhaul and modernisation of mine safety legislation culminating in the passage of the Mines Safety and Inspection Act 1994. The Act, which was introduced and passed by Parliament in 1994, brought the administration of mine safety in all types of mines under a single administrative body and piece of legislation, repealing the Coal Mines Regulation Act 1946 and Mines Regulation Act 1946. The Mines Safety and Inspection Act 1994 received Royal Assent on 7 November 1994 and commenced in December 1995.

A joint approach to the introduction of Robens-type Coal Mines Regulation Act 1946 and Mines Regulation Act 1946. The Mines Safety and Inspection Act 1994 received Royal Assent on 7 November 1994 and commenced in December 1995. Principles to improve standards of occupational health and safety, which included the involvement of the contracting sector, began to impact on industry safety performance. The supportive response to the concerns of contractors led to the establishment of the contractor safety training organisation the Mining and Resource Contractors Safety Training Association (MARCSTA), which was to play a significant role in the delivery of high-quality safety inductions to most sectors of the mining industry.

This not-for-profit organisation, formed in 1994 and incorporated in 1996, consisted of 19 operating companies* whose primary objective was to replace repetitive and other questionable induction programs which were not providing an appropriate standard of induction training and were resulting in a negative response to occupational health and safety per se and incurring significant cost consequences. The formation of MARCSTA was given strong and enthusiastic support by the State Mining Engineer and his staff and was endorsed by the industry body.

From the year 1989–90 the occupational safety and health performance of the industry began to reflect the combined efforts of industry, its contractors and the regulatory authority, and the industry began gradually to move toward its target of world leadership in occupational health and safety. By 1998 the membership of the Chamber of Mines included no less than 17 major contractors, many of whom were represented on the various occupational health and safety committees of the chamber, particularly in the Eastern Goldfields region.

The outcome of contractors' occupational safety and health management

Contractors by this time were becoming increasingly active in all sectors of the industry and were demonstrating their commitment to achieving high standards of health and safety to meet industry's expectations. The incidence of fatalities (number per 1,000 employees) declined from 0.527 in 1988–89 to 0.047 in 2010–11. (See Figures 6.3 and 6.4.)

The incidence of lost time and serious injuries for Western Australian mining industry workers was reduced similarly.

* Atlas Copco, AWP Contractors, BGC, Boral, Brandrill, Charles Hull, Clough Engineering, CSR Readymix, Eltin, Henry and Walker, JR Engineering, McMahon, Minepro, Monadelphous, NS Komatsu, Roche Brothers, Skilled Engineering, Thiess, Total Corrosion Control.

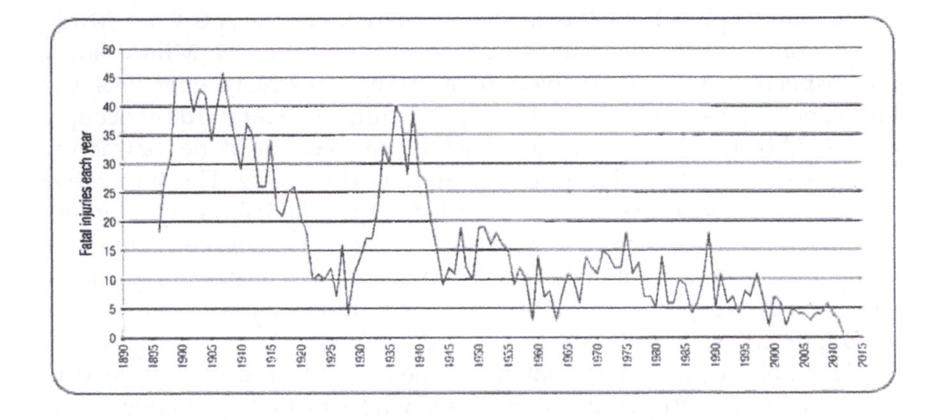

Figure 6.3 Fatal incidence rate per 1000 employees. (From Department of Mines and Petroleum. (2013). In 2012 there were no fatalities in the Western Australian Industry (Kerr 2013). The incidence of lost time and serious injurious for the Western Australian mining Industry workers was reduced similarly.)

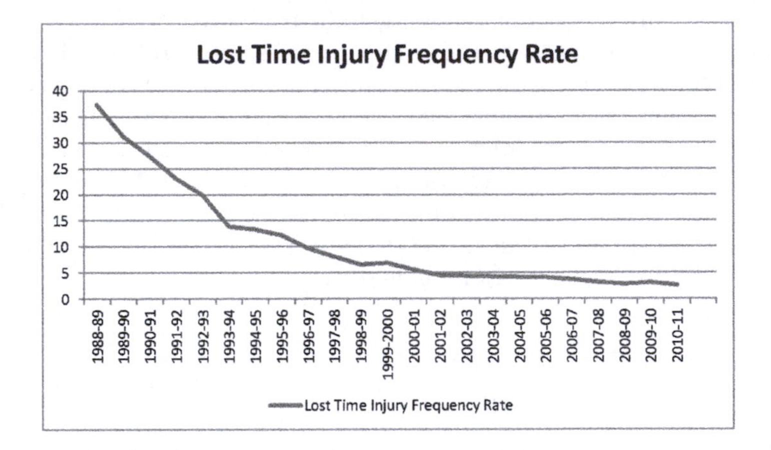

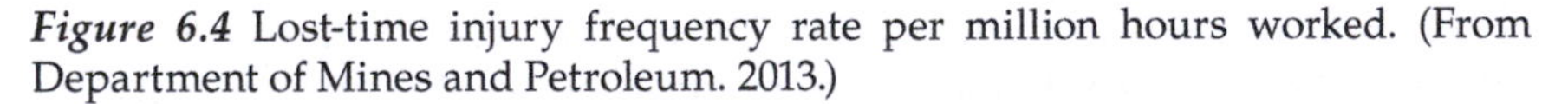

Figure 6.4 Lost-time injury frequency rate per million hours worked. (From Department of Mines and Petroleum. 2013.)

In 2013 contractors now provide the majority of production employees and are respected for the high standards of health and safety practices by their workforces.

Workers' compensation premiums are perhaps the most objective indicator of the health and safety standards existing in major industry sectors. The continual downward trend in premiums for the mining sector, particularly from 1990 onward, provides convincing and irrefutable evidence of the industry's status as a world leader in occupational health and safety.

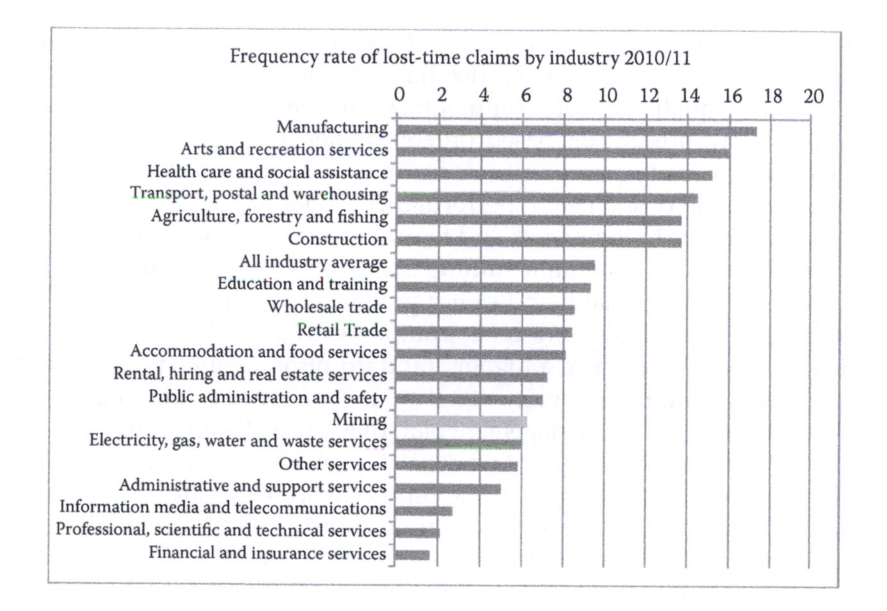

Figure 6.5 Frequency rate of lost-time claims by industry, 2010–11. (From Western Australian Department of Commerce. (2012). *WA Key OSH Statistics*.)

The contracting industry can rightly claim to have been a major influence in achieving this status and perhaps the most noteworthy aspect is the total absence of any conflict with major mining companies over the past 25 years in the transition occasioned by the introduction of the Robens-style occupational health and safety legislation in Western Australia.

The Western Australian Department of Commerce (2012) has released data on occupational safety and health in Western Australia derived from workers' compensation claims of workers covered by the Workers' Compensation and Injury Management Act in 2010–11. (See Figure 6.5.)

It is clear from this information that the mining industry in Western Australia is performing at a level unmatched in the world today.

Conclusions

Unlike in many industries that mainly have company employees, in Western Australia in the mining industry, contractors form the major portion of the workforce. Significant improvement in health and safety practices in the Western Australian mining industry has been due to the work of contractors. Today, Western Australian contractor mining industry workers are educated concerning important occupational health and safety legislation requirements, about working safely and maintaining healthy work practices in Western Australia, due to the induction and ongoing education provided by MARCSTA. Contractor companies working in the mining industry also have developed and use integrated health, safety, quality and environmental management systems and plans, standard work procedures based on comprehensive job safety analysis, have personal safety plans and use leading and lag indicators to assess occupational safety and health performance to identify any opportunities for improvements in their work and to make these improvements.

Contractors work for many mining companies. From the mining companies with the best health and safety practice they learn to meet the set standards of work. There are other companies that contractors work for where the contractors have a higher standard of occupational health and safety than the mining company that they are employed to work for, and it is the contractor's safety management plan that is used throughout the work site to deliver a high standard of work-related health and safety practices.

An outcome of having a high standard of workplace safety and health practices by contractor and other mining industry employees is demonstrated in this chapter through the use of the figures that show Western Australian mining employees provide the highest level of profit per hour worked (Figure 6.2), have decreased employee fatal injury rate (Figure 6.3), decreased lost-time injury rate (Figure 6.4) and by the fact that the mining industry in Western Australia has a lower frequency rate of lost-time workers' compensation claims than manufacturing, arts and recreation services, health care and social assistance, transport, postal, warehousing, agriculture, forestry, fishing, construction, education and training, wholesale trade, retail trade, accommodation and food services, rental, hiring and real estate services and public administration employees. In Western Australia the mining industry has become one of the safest industries to work in.

Acknowledgement

Paula Sinclair is thanked for her work in proofreading this chapter.

References

Australian Mining History. (2012). Retrieved from http://www.readyed.com.au/Sites/minehist.htm.

AXTAT Incident Reports Data Base. Retrieved from http://search.doir.wa.gov.au/search?q = AXTAT&site = search_dmp&client = search_dmp&output = xml_no_dtd&proxystylesheet = search_dmp&filter = 0&as_dt = 1; collation of AXTAT data base figures.

Department of Mines. (1981). *Western Australian Annual Report 1981*. Perth, WA: Government of Western Australia.

Department of Mines. (1989). *Western Australia 1988–1989 Yearbook*. Perth, WA: Government of Western Australia.

Department of Mines and Petroleum. (2012a, March, 31). *Mining and Petroleum Investment*. Retrieved from http://www.dmp.wa.gov.au/12410.aspx.

Department of Mines and Petroleum. (2012b). *Western Australia's Petroleum and Geothermal Explorer's Guide*. Perth, WA: Government of Western Australia.

Department of Mines and Petroleum. (2013). *AXTAT Incident Reports Data Base*. Retrieved from http://search.doir.wa.gov.au/search?q = AXTAT&site = search_dmp&client = search_dmp&output = xml_no_dtd&proxystylesheet = search_dmp&filter = 0&as_dt = 1.

Kerr, P. (2013) 116 year milestone. Finally a year when no one died. The West Australian–West Business. p.1.

Mineral Resources. (2007). *Mining by Aborigines—Australia's First Miners*. New South Wales Department of Primary Industries. Retrieved from http://www.dpi.nsw.gov.au/__data/assets/pdf_file/0008/109817/mining-by-aborigines.pdf.

Resources Industry Training Council. (2010). *Western Australian Mining Industry: Workforce Development Plan*. Perth, WA: Author.

The Chamber of Minerals and Energy of Western Australia. (1997). *A Guide to Contractor Occupational Health and Safety Management for Western Australian Mines*. Perth, WA: Author.

The Chamber of Minerals and Energy of Western Australia. (2012a). *Diversity in the Western Australian Resources Sector Survey*. Perth, WA: Author.

The Chamber of Minerals and Energy of Western Australia. (2012b, November). *WA State Growth Outlook 2013*. Perth, WA: Author.

The Chamber of Mines of Western Australia [Inc.]. (1987). *86th Annual Report, 1987*. Perth, WA: Author.

Western Australian Department of Commerce. (2012). *WA Key OSH Statistics*. Retrieved from http://www.commerce.wa.gov.au/worksafe/pdf/statistics-industry/wa_key_osh_stats.pdf. p. 60.

chapter seven

Safety duties and reliance on expert contractors

The Cyclone George disaster

Olga Klimczak
LLB (Hons.) BA (Hons.)
LLM Candidate, University of Melbourne

Contents

> The circumstances surrounding the building and installation of the dongas at RV1 and the events that unfolded on 9 March 2007 has brought into scrutiny the measures taken by companies and persons to safely protect employees and property from the effects of a cyclone such as TC George. (*Department of Consumer and Employment Protection v The Pilbara Infrastructure Pty Ltd & Anor*, 2011, [18])

Introduction

In the early hours of the morning on 9 March 2007, severe category 4 Tropical Cyclone George passed over a camp comprised of temporary prefabricated residential accommodation units (commonly known as *dongas*) and other facilities. The camp, known as Rail Village 1 (RV1), was located about 90 kilometres from the coast near Port Hedland, in the Pilbara, Western Australia. It predominantly accommodated workers (both construction and management) working on the construction of a private railway between Fortescue Metals Group's (FMG) Cloudbreak iron ore mine and its port facilities at Port Hedland. Workers servicing and maintaining the dongas and other facilities were also accommodated at RV1.

The workers had taken refuge in the dongas to shelter from Cyclone George. This was common practice in the Pilbara. But, unfortunately, some of the dongas were not safe refuge. The cyclonic winds of Cyclone George caused these dongas to become dislodged from their concrete footings, and others to lift and pull away from their tie downs. Some dongas flipped and/or collided with other dongas. Two workers died as a result of the injuries they sustained from the effects of Cyclone George and many were severely injured.

In the aftermath, investigations revealed that a number of failures contributed to the disaster, including that the dongas had not been designed to withstand the level of cyclonic winds set by the Australian Standards for the region in which they were located, that the design for the tie downs securing the dongas to their concrete footings was deficient, and that the welding securing the dongas to the tie downs had been carried out by an unqualified welder and were of poor quality (*Department of Consumer and Employment Protection v The Pilbara Infrastructure Pty Ltd & Anor*, 2011, [12], [19]).

Like many disasters, the tragic and severe consequences of Cyclone George resulted in a spotlight being shone on the conduct of the employers and principals of the workers who had been injured or killed. Had they discharged their duties under applicable occupational health and safety legislation? That is, had they, so far as was reasonably practicable, provided and maintained a working environment in which employees and contractors were not exposed to hazards?

One of their answers to this question was that they had discharged their duty by relying on the specialist expertise of others in the design and construction of the dongas, and that it was not reasonably practicable to take further steps under the circumstances (*Department of Consumer and Employment Protection v The Pilbara Infrastructure Pty Ltd & Anor*, 2011, [68]–[73]). However, it also appeared that multiple parties assumed that someone else would ensure that the wind region specifications of the dongas were correct, and the error in the initial request for tender documents for the supply of the dongas was not corrected by anybody at any stage of the process. That gap had catastrophic consequences.

The issue

The use of expert contractors to perform certain specialised work for a party is not a new phenomenon. With the advent of technologies, expansion of industries, the increased scale of projects and the development of new products, plant, and systems, it is more and more commonplace for parties to engage expert contractors to perform work where the party does not itself have relevant expertise. Indeed, if a party does not have expertise in relation to a potential hazard at the workplace to which its employees or others may be exposed, it must bring that expertise to bear on the task (*Hamersley Iron Pty Ltd v Robertson*, 1998). Further, commercial work arrangements are increasingly becoming more complex, with multiple layers of subcontractors often being engaged. On major construction projects, there are usually numerous parties involved in various aspects of the work associated with the project. The intersection of these various work relationships raises questions about how best to manage occupational health and safety issues, and the scope of the duty of each party.

One of these questions is, to what extent must parties engaging expert contractors themselves take additional steps (rather than simply engaging the expert contractor), in order to discharge their safety obligations and to effectively manage contractor safety issues? For example, do these parties have to supervise the expert contractor's work, or engage additional experts to do so? Should they give directions to the expert contractors about safety matters? Is this feasible given their limited knowledge and lack of expertise? What are the limits of reliance on an expert contractor's expertise, if any? These questions have been considered in a number of recent Australian cases (see Chapter 3), including the prosecution of the employers and principals whose workers were injured in the Cyclone George disaster (*Department of Consumer and Employment Protection v The Pilbara Infrastructure Pty Ltd & Anor*, 2011; *Kirwin v The Pilbara Infrastructure Pty Ltd*, 2012. *Kirwin v Laing O'Rourke (BMC) Pty Ltd*, 2009; *Kirwin v Laing O'Rourke (BMC) Pty Ltd*, 2010. *Laing O'Rourke (BMC) Pty Ltd v Kirwin*, 2011).

This chapter will consider the safety issues that arise in the context of the engagement and reliance on expert contractors, including the legal limits on such reliance in the context of a party's safety duty. Using the Cyclone George disaster as a focal case study, it will argue that parties must adopt a proactive and comprehensive approach to contractor safety management at all stages of the contracting process, together with a collaborative approach between principal and contractor, so as to try to achieve the best possible outcome for workers and others who may be exposed to health and safety risks and hazards in the work context. It will also contend that it is vital that parties clearly communicate about what aspects of work they are relying on expert contractors for, or are assuming they will perform, to ensure that the expert contractors are aware of this.

This is not to say that such an approach will identify safety issues and gaps in every case, or will ensure tragic workplace accidents or deaths never occur. However, by duty holders engaging proactively in a dialogue with other relevant parties, including expert contractors, about the reliance that is being placed on that other party for the identification of any risks and control measures that are outside the expertise of the first party, there is a greater chance that issues or gaps may be identified at an early stage and be remedied.

The legislative framework

It is useful to first briefly explain the legislative context in which the safety duties arise, so as to provide a framework for the discussion in this chapter.

Earlier chapters have set out the occupational health and safety (OSH) regimes that apply in Australia, so it is not proposed to go into this in detail in this chapter.* The cases discussed in this chapter in relation to the Cyclone George disaster were subject to the West Australian OSH legislative framework. Therefore, the key elements of that framework, which are relevant to the issues discussed in this chapter, will be highlighted. Further, this chapter focuses on the safety duties of employers and principals, given the parties who were prosecuted were employers and/or principals.

Under the West Australian OSH laws, the primary duty holder is the employer, who, broadly, has a safety duty in respect of its employees and others who may be adversely affected by the employer's work (Occupational Safety & Health Act 1984 (WA), ss 19[1], 21[2]). Principals are also duty holders and have a safety duty, in relation to matters over which they have the capacity to exercise control, to their contractor's employees (Occupational Safety & Health Act, s 23D).

In terms of the nature of the duty, the first key issue is that the duty is personal and non-delegable (*Department of Consumer and Employment Protection v The Pilbara Infrastructure Pty Ltd & Anor*, 2011, [102]). This means that liability for the duty cannot be transferred or delegated to another party (*Kondis v State Transport Authority*, 1984, pp. 679, 681). However, a party can engage a third party, such as an expert contractor with specialist expertise, to assist it in discharging its duty (*Kirwin v The Pilbara Infrastructure*

* Chapter 3 also discusses the key features of the model workplace health and safety legislation that are relevant to this issue. The model legislation has been adopted by most Australian states and territories, but not Western Australia or Victoria at the time of writing this chapter.

Pty Ltd, 2012, [180]). Duties may be held concurrently by multiple duty holders and may overlap in scope.

The other fundamental aspect of the duty is that it is not absolute; it is limited by what is reasonably practicable (Occupational Safety & Health Acts 19(1)). The obligation on the duty holder is to ensure, *so far as is practicable*, the safety and health of those to whom the duty is owed. The term *practicable* is defined to mean reasonably practicable and having regard for, where the context permits:

(a) the severity of any potential injury or harm to health that may be involved, and the degree of risk of it occurring; and
(b) the state of knowledge about the
  (i) injury or harm to health in (a);
  (ii) risk of that injury or harm to health occurring; and
  (iii) means of removing or mitigating the risk or the potential injury or harm to health; and
(c) the availability, suitability, and cost of the means referred to in paragraph (b)(iii)
(Occupational Safety & Health Act, s 3)

The Australian courts' approach to determining whether a party has met its safety obligations is a practical one; it involves looking at the facts of each case without the benefit of hindsight, or the wisdom of Solomon, but recognising the possibility of human inadvertence or spontaneous stupidity, and the importance of a duty holder's safety duty (*Holmes v R E Spence & Co Pty Ltd*, 1993). Duty holders are expected to take a proactive, imaginative and flexible approach to their safety obligations (*Holmes v R E Spence & Co Pty Ltd*, 1993). This includes proactive contractor safety management.

Engaging expert contractors

Applying the above legal framework to the issues in this chapter (the engagement of expert contractors), the key issue is what are the reasonably practicable steps that a duty holder must take to minimise potential risks and hazards? The Cyclone George cases discussed later demonstrate that the courts have engaged in a balancing exercise in determining what steps are reasonably practicable for a party to take in relation to an expert contractor's work (*Laing O'Rourke (BMC) Pty Ltd v Kirwin*, 2011, [31], [35]. *Kirwin v Laing O'Rourke (BMC) Pty* Ltd, 2010, [95]–[97]). In some cases, the engagement of an expert contractor will, of itself, be sufficient to discharge the duty holder's safety duty, as it will not be reasonably practicable for the duty holder to do anything further (*Laing O'Rourke (BMC) Pty Ltd v Kirwin*, 2011, [5], [67]. *Kirwin v Laing O'Rourke [BMC] Pty*

Ltd, 2010, [149]–[150], [158]). In other cases, there may be other reasonably practicable measures for the duty holder to take to discharge its safety duties (*Hamersley Iron Pty Ltd v Robertson*, 1998). However, each case will depend on the surrounding circumstances and an assessment of the potential steps that might have been taken in relation to the potential hazards.

Ultimately, the Cyclone George cases turned on the reasonableness of the duty holders' assumptions in the context of assessing what it was reasonably practicable for the duty holders to do. Both the court of appeal and Hall J were not prepared to find that it was reasonably practicable for the relevant duty holders to engage *additional* expert contractors to supervise or check the work of first-expert contractors engaged by the duty holder, as there was nothing to put them on notice that it was not reasonable to rely on the first-expert contractor (*Laing O'Rourke (BMC) Pty Ltd v Kirwin*, 2011, [5], [67]–[68]. *Kirwin v Laing O'Rourke (BMC) Pty* Ltd, 2010, [149]–[150], [158], [181]–[183]). Therefore, if the party engaging the expert contractor is, or should reasonably be, aware of any issue in relation to the expert contractor's competence or their work, it may not be reasonable for the party to rely on the expert contractor in discharging its safety duties, and there may be other reasonably practicable steps for it to take. Further, as other cases have made clear, parties still need to be vigilant in relation to safety matters that are within their control or scope of expertise (*R v ACR Roofing*, 2004. *Baiada Poultry Pty Ltd v The Queen*, 2012).

Cyclone George disaster as a case study

Introduction

The Pilbara region in Western Australia is susceptible to tropical cyclones from around November to April each year. Cyclones are notoriously unpredictable, and this was demonstrated by the abruptness with which Cyclone George changed its course and hit the coast near Port Hedland and then travelled inland over RV1.

By the time Cyclone George passed over RV1, in the early hours of the morning, all workers at RV1 were in the dongas allocated to them. This was consistent with the cyclone procedure, which provided that workers were to take refuge in their dongas during a red alert, including on impact.

It took this cyclone to make obvious the need for adequate building guidelines to be followed to avoid a safety hazard to the workers at RV1 during its high winds and rain. The risk that had been taken was made plain when some of the dongas did not withstand the cyclonic winds. The scenes of devastation and destruction of these dongas at RV1 following Cyclone George told the haunting story of what the workers had been exposed to—parts of roofing, walls, mattresses and other debris were strewn across the site. Tragically, some of the workers who had been staying in the destroyed dongas were injured or killed. As luck would have it, other dongas survived the impact of Cyclone George intact and with no injury to the workers who had been sheltering in them.

The prosecutions

In the aftermath of the Cyclone George disaster, the regulator, WorkSafe WA (WorkSafe), initiated prosecutions against a number of parties. Given the complexity of the contractual arrangements and the number of parties involved, it is perhaps not surprising that this scattergun approach was adopted, at least initially. However, what is surprising is that, ultimately, the prosecutions that were commenced were almost all either unsuccessful or discontinued. This is notwithstanding that the evidence clearly established that errors had been made in the design, construction, and installation process. This may come down to the actions of WorkSafe during the course of prosecuting the charges.* However, it also demonstrates that the safety duty under the OSH regimes is not absolute. The role of the court is to determine whether the accused were criminally responsible for having committed certain alleged offences as set out in the prosecution (*Department of Consumer and Employment Protection v The Pilbara Infrastructure Pty Ltd & Anor*, 2011, [21]). As Magistrate Mignacca-Randazzo stated (*Department of Consumer and Employment Protection v The Pilbara Infrastructure Pty Ltd & Anor*, 2011, [20]):

> The tragic and devastating consequences brought about [by Cyclone George] no doubt had and will continue to have profound and long-lasting effects on the lives of so many victims directly and indirectly affected by what happened at RV1 that night. . . . Whilst this Court recognises that state of affairs, nothing stated here or done in the course of these proceedings will ever materially change what happened or bring sufficient (if any) relief to those many victims.

* Magistrate Mignacca-Randazzo heavily criticised NT Link and Mr Lawry. Worksafe had, at an earlier stage, agreed not to lead any evidence in the prosecution of NT Link, and the charges against NT Link were therefore dismissed. Mr Smith (a director of NT Link), his wife, and the parent company of NT Link were granted immunity, and Mr Smith testified against FMG and TPI. During the course of the trial, Mr Smith was found to have, without permission, made notes on his hand and read from them when giving his evidence: his evidence was found to be unsatisfactory in relation to some issues.

The parties who were prosecuted by WorkSafe broadly fell into two categories:

a. Those who had some involvement in the construction process of RV1.
b. Those whose employees or contractors were staying at RV1 at the time of Cyclone George, and were injured or killed.

There were two sets of reported decisions (including appeals from the initial magistrate's decision): the prosecution of FMG and its wholly owned subsidiary, The Pilbara Infrastructure Pty Ltd (TPI), and the prosecution of Laing O'Rourke (BMC) Pty Ltd (Laing) (*Department of Consumer and Employment Protection v The Pilbara Infrastructure Pty Ltd & Anor*, 2011; *Kirwin v The Pilbara Infrastructure Pty Ltd*, 2012. *Kirwin v Laing O'Rourke (BMC) Pty Ltd*, 2009; *Kirwin v Laing O'Rourke [BMC] Pty Ltd*, 2010. *Laing O'Rourke [BMC] Pty Ltd v Kirwin*, 2011). The facts and background set out in this chapter are drawn from these decisions.

FMG and TPI fell within both the first and second categories referred to above. Some of FMG and TPI's contractors (including where the contractor's employer also had charges laid against them) were injured or killed when Cyclone George hit their dongas at RV1. Additionally, both FMG and TPI were involved in the design, construction and installation process of the dongas, although, as will be seen below, they engaged other expert contractors to perform this work.

Laing was a contractor of FMG engaged to perform railway track and bridge work on the construction of the private railway line linking FMG's Cloudbreak Mine to port facilities at Port Hedland. Laing had both contractors and employees working on the project, all of whom were accommodated in the dongas at RV1, and some of whom were injured or killed during the Cyclone George disaster. Whilst its workers were accommodated at RV1, Laing had no involvement whatsoever in the construction of RV1. Therefore, it fell within the second of the categories referred to above.

Charges

The charges laid were broadly similar, and related to the alleged failure by the accused to take reasonable steps to minimise the risks associated with the exposure of its employees and contractors to the hazard of Cyclone George. Specifically, the steps alleged (*Department of Consumer and Employment Protection v The Pilbara Infrastructure Pty Ltd & Anor, 2011. Kirwin v Laing O'Rourke (BMC) Pty Ltd,* 2010) to have been reasonably practicable but were not taken include

a. having in place proper cyclone procedures (that included both evacuation to a safe refuge at RV1 or elsewhere if there was no safe refuge at RV1);

b. training workers in the cyclone procedures;
c. ensuring that the dongas were safe refuge; and/or
d. evacuating workers if no safe refuge was available at RV1.

This chapter focuses on the safe refuge charges in paragraph (c) above. It was clear that the dongas were intended to be used as safe refuge and that some were not safe refuge when Cyclone George passed over RV1. The main issue in the prosecutions was the extent to which parties could rely on certain facts and the engagement of expert contractors to assume the dongas were safe refuge, without taking any additional steps to ensure or check the dongas were in fact suitable to be used as safe refuge in the event of a cyclone.

Outcomes

At trial, FMG, TPI and Laing were all acquitted, and ultimately successful in relation to the appeals that followed. However, it should be noted that in respect to Laing, Murray J set aside the acquittal and convicted Laing of the charges, but that this was reversed by the court of appeal. When the convictions against Laing were quashed by the court of appeal, WorkSafe discontinued the prosecutions against other contractors in a similar position (Campbell, 2010).

Key facts

The facts leading up to the Cyclone George disaster are complex and, to some extent, may have contributed to the assumption by all parties that someone else had ensured the design, construction and installation of the dongas was competent and consistent with Australian Standards. In particular, it appears that all parties seemed to rely on another party for ensuring the dongas were built in accordance with the correct wind specification for the proposed location, so as to ensure they were safe refuge in the event of a cyclone in the area.

FMG owned the Cloudbreak Mine, an iron ore mine located about 250 km south of Port Hedland. Iron ore from the mine was to be exported using port facilities at Port Hedland. To facilitate this, FMG decided to build a private railway and associated infrastructure for the purpose of transporting the iron ore from the Cloudbreak Mine to the port facilities. FMG/TPI engaged WorleyParsons to carry out project management services, including engineering design services, procurement services, contract development and construction management services (EPCM). FMG/TPI also engaged other contractors to perform various works associated with the construction of the railway, including Laing and others against whom charges were laid by WorkSafe. Given the remote location of the work on the railway line, two non-permanent accommodation facilities (construction camps) were to be built to accommodate the workers. One of these camps was RV1.

RV1 was to be built about 90 km from the smooth coastline near Port Hedland. This meant that, according to the Australian Standards, it was located in cyclonic Wind Region C, being the region between 50 and 100 km from the smooth coastline. The area was prone to cyclonic winds between November and April. The other wind regions in the Australian Standards are Wind Region D (located from the smooth coastline to 50 km inland, and is the area associated with the most severe cyclonic winds), Wind Region B (located between 100 to 150 km from the smooth coastline) and Wind Region A (located more than 150 km from the smooth coastline). Wind Regions A and B are not cyclonic wind regions.

The design strength of a structure set out in the Australian Standards is broadly determined on the basis of the likely wind speed and the probabilities of this wind speed being exceeded. In the Australian Standards, the difference between the wind speed that a residential building must be built to withstand in Wind Region C compared to Wind Region A is substantial—approximately 90 km/hr (about 250 km/hour compared with 162 km/hour). This turned out to be significant, because, during the supply process, an error was made in identifying the relevant wind region that applied to the location of RV1.

FMG/TPI initially engaged Spotless Services Pty Ltd (Spotless) as its project manager in relation to the accommodation facilities at RV1. Its role was twofold. First, on behalf of FMG/TPI, procure the design, finance, manufacture, construction, transportation, installation and commissioning (together referred to as the 'supply') of the accommodation facilities (i.e., the dongas) at RV1. This included assisting with obtaining the necessary building approvals from the relevant shire. It was intended that Spotless would, on behalf of FMG/TPI, engage a contractor to supply the accommodation facilities. Second, once supplied, Spotless would operate and manage the facilities.

Spotless undertook a tender process for the supply of the dongas. As part of this process, Mr Guthrie (Spotless' construction manager) and others attended the proposed site for a physical site inspection. Spotless (primarily by Mr Guthrie) then prepared a request for tender (RFT), which included certain design specifications and technical requirements. Critically, the RFT specifications incorrectly identified Wind Region A as the applicable wind region (this had been determined by Mr Guthrie on behalf of Spotless). This error was then mistakenly replicated in all other subsequent documentation related to the supply of the dongas. The RFT was circulated to FMG/TPI representatives for their sign-off. Spotless then completed the tender process by seeking tenders, providing the RFT and reviewing tenders received from potential contractors. However, no acceptable tenders were received (based on price).

Spotless representatives then identified Spunbrood Pty Ltd, trading as NT Link (NT Link), who had not tendered in the initial process, as a potential supplier of the dongas. With FMG/TPI's approval, Spotless proceeded to negotiate with NT Link for the supply of the dongas by NT Link. NT Link was provided with a copy of the RFT documentation and submitted a quote for the contract. However, the negotiations between Spotless and NT Link broke down. At this point, Spotless withdrew from the procurement process and solely agreed to operate and manage the facilities upon their completion. However, FMG/TPI's understanding was that Mr Guthrie would continue to be involved in the supply of the dongas on behalf of NT Link.

TPI now stepped in and began dealing directly with NT Link. The terms of the supply were agreed and TPI engaged NT Link to supply the accommodation facilities under a written contract. The contractual arrangements included that the dongas would remain the property of NT Link's parent company, Smith Prell Pty Ltd, and would be leased to TPI during the construction work on FMG's railway. As temporary accommodation units, they could be easily dismantled and removed once they were no longer required.

Prior to the installation of the dongas, the Shire of East Pilbara approved the plans, designs and diagrams specifying certain specifications and design criteria by which the dongas would be built. The application for planning approval had been lodged at an earlier stage by Spotless, and was signed by both a representative of Spotless and of FMG/TPI. At a later stage, prior to commencing installation of the dongas, NT Link lodged an application for a building licence with the shire. The application included plans and diagrams, some of which had on them a stamp with the words 'Certification Robin Salter & Associates Chartered Consulting Engineers', which was signed (but no person from that engineering firm gave evidence in the prosecutions). Despite the clearly incorrect wind specification in the application (Wind Region A and not Wind Region C), the shire approved the plans and issued a building licence to NT Link. However, a condition of approval was that the construction was to be carried out in accordance with the Building Code of Australia, which included requirements in relation to wind regions and related specifications. The magistrate in the FMG/TPI prosecution found the dongas did not comply with the requirements for wind region A and certainly not Wind Region C (*Department of Consumer and Employment Protection v The Pilbara Infrastructure Pty Ltd & Anor*, 2011, [119]).

NT Link then proceeded to complete the supply process of the dongas. This included the design of the tie downs for the dongas, which was carried out by Mr Guthrie on behalf of NT Link. Despite earlier representations by NT Link that it would engage consultant engineers to assist it with the supply of the dongas, the design of the tie downs was not approved by any engineers. The magistrate in the FMG/TPI prosecution found the tie down design lacked horizontal restraint, and was deficient and inappropriate for the location of RV1 in Wind Region C, and also inadequate for Wind Region A (*Department of Consumer and Employment Protection v The Pilbara Infrastructure Pty Ltd & Anor*, 2011, [19], [302]–[303]). However, this deficiency was not identified by anyone until after the Cyclone George tragedy. NT Link also transported and installed the dongas at the RV1 site. This included preparing concrete footings and welding the tie downs of the dongas to the concrete footings.

TPI had engaged Mr Lawry, an engineer engaged by WorleyParsons, to supervise the installation and commissioning of the dongas at the RV1 site by NT Link, and ensure quality control. During the Cyclone George cases a number of issues with the installation of the dongas were identified. These were: first, different tie downs were used than that specified in the shire-approved plans. Although they were of a higher standard than the design, they remained deficient in material respects. Second, the welder performing the welding work joining the tie downs to the flooring of the dongas was unqualified and the welds were defective. Other aspects of the work were also substandard. Mr Lawry did not identify these issues, and nothing in his reports indicated anything other than that the dongas had been installed according to plans approved by the shire in a competent manner.

Once completed, the dongas began to be used as residential accommodation units for railway construction and related workers until the disaster. Workers were allocated to particular dongas by a team comprising FMG, TPI and Worley Parsons representatives known as 'Team 45'.

Key issues

Clearly, the task of designing and supplying the dongas required expert skill, particularly in light of the fact that the dongas were to be installed in a cyclonic region and used as shelters in the event of a cyclone.

Neither FMG nor TPI's key representatives involved in the supply process, nor anyone else at FMG or TPI, had any relevant expertise in relation to the supply of dongas in cyclonic regions. FMG/TPI's answer to the charges laid against them was that they had conducted their affairs and entered into contracts with experts to perform the specialist work and relied on those expert contractors to meet the necessary standards. FMG and TPI argued that there was nothing else reasonably practicable for them to do (*Department of Consumer and Employment Protection v The Pilbara Infrastructure Pty Ltd & Anor*, 2011, [68]–[73]).

Similarly, Laing relied on other parties in relation to the question of whether the dongas were safe refuge in the event of a cyclone. It did not engage anyone to inspect the dongas to ensure they were safe refuge, nor did it seek any assurances to that effect from FMG/TPI or WorleyParsons (*Kirwin v Laing O'Rourke (BMC) Pty Ltd*, 2009, pp. 90–91).

The reliance of each of these parties is discussed below.

Reliance by FMG/TPI on expert contractors

The expert contractors FMG/TPI engaged and relied on to assist it with the supply of the dongas for RV1 were Spotless, NT Link and WorleyParsons. The following factors were relevant to that reliance:

a. the expert contractors' representations and FMG/TPI's belief as to their competence and expertise;
b. the contractual documentation, which dealt with the allocation of responsibility;
c. the fact that the expert contractors' appeared to FMG/TPI to undertake the work competently; and
d. the lack of expertise and control by FMG/TPI.

1. Competence and expertise of expert contractors. Spotless, NT Link and WorleyParsons (Mr Lawry) each represented that they were competent, and had the relevant expertise, qualifications and capacity to complete the works required of them. FMG/TPI's representatives believed Spotless was a strong credible organisation and was capable of delivering the accommodation facilities within budget. Spotless represented to FMG/TPI that it had the resources, capacity and capability to do so and that senior managers would be closely involved. Spotless was not involved in the resources sector at the time it was engaged by TPI. However, Mr Guthrie, was a registered builder and construction manager, had drafting and designing skills, and had experience working in the Pilbara and in the design and installation of buildings, including camp accommodation at remote mining locations. Spotless also appointed a firm as consultant architects and engineers to assist it in the supply process, as and when the need arose. FMG/TPI's representatives were impressed with and relied on the experience and credentials of Spotless, including in particular, Mr Guthrie, who had relevant experience and expertise in camp design and construction.

In relation to NT Link, when the tender process failed, Spotless representatives informed FMG/TPI's representatives of Mr Smith at NT Link, and that Mr Smith had previously built camps in remote areas and was reliable. NT Link regarded itself as a substantial, well-structured and managed company, professing significant expertise and experience in the design and supply of transportable buildings, such as camp buildings to be installed in cyclonic regions. The RFT was provided to Mr Smith and used by NT Link as the basis for a written quote submitted to TPI. The quote conveyed that, relevantly, the dongas would be built by qualified persons in accordance with Australian Standards, including the Building Code of Australia, and that the tie downs would be for Wind Region A and would be designed by an engineering firm engaged by NT Link for that purpose. NT Link represented to TPI that it had substantial experience in the design and installation of railway camps, having recently completed such camps for the Alice Springs to Darwin railway. NT Link also represented (by Mr Smith to FMG/TPI's representatives) that NT Link had experience in the installation of cyclonic work camps and capacity to complete the work. Additionally, FMG/TPI's representatives also understood that Mr Guthrie (who had been involved in the RFT and design work on behalf of Spotless) would continue to be involved on behalf of NT Link after Spotless withdrew from the process. As NT Link did not have any engineering capacity, it conveyed that it had access to engineering services and expertise. Some of the drawings annexed to the installation contract had what appeared to be a certification stamp by a firm of chartered consulting engineers with the word 'Certified' and a signature of what appeared to be an engineer.

Finally, Mr Lawry was a civil engineer with about 35 years of experience at the time of the supply of the dongas. He had engineering qualifications and mining experience, and was engaged by WorleyParsons, the EPCM. FMG/TPI representatives considered him to be a 'competent operator'.

2. *Contractual documentation and allocation of responsibility.* The contractual arrangements and documentation made the expert contractor (i.e., NT Link) responsible for the design, fabrication, manufacture, supply, installation and provision of services for RV1. FMG/TPI were not responsible for, and did not attempt to check or supervise, the technical aspects of the supply process, including in particular, the design. FMG/TPI representatives saw their role as agreeing pricing and ensuring the amenities and facilities were as required by FMG.

Initially, it was Spotless' responsibility (under an MOU with FMG) to finalise the detailed design specification, and to detail the specifications and technical requirements for the supply process. These were set out in the RFT. In turn, The RFT provided the following in relation to the allocation of responsibility between the parties:

- The tenderer was to inform itself, prior to submitting a tender, including by (1) becoming acquainted with the nature of the project, the work to be performed on the project, and the proposed risk allocation set out in the tender documents; and (2) by inspecting the site and/or local conditions affecting the performance of the works, including the nature and location of the site. There was also a disclaimer in relation to there being no warranty or representations as to the completeness or accuracy of the information in the RFT. By submitting a tender, the tenderer warranted that it had relied on its own inquiries and not the Tender Documents (being the RFT and its annexures).

- The works were to be in accordance with the specifications in the RFT, and the detailed design was to conform with the law and specifically, the Australian Standards and Building Code of Australia. Layout designs were to be similar to those provided by Spotless. The tenderer was responsible for ensuring the camps were designed and built competently and in accordance with all relevant standards.
- The tenderer was to carry out the design with all the care and skill to be expected of appropriately qualified and experienced professional designers and engineers with experience in carrying out the works. The tenderer was to be fully responsible for all pre-contract design, whether or not carried out by it. Although all design documentation was to be submitted to Spotless as the Project Manager for approval, full design responsibility rested with the tenderer.

The RFT therefore clearly contemplated that suitably qualified and experienced personnel of the supply contractor would carry out and be responsible for the design. At the same time, it was also envisaged that Spotless would check the design proposed by the tenderer, which was to comply with the specifications in the RFT, and regulatory and legal requirements. The specifications in the RFT included that design wind loads related to Wind Region A. Significantly, due to the proposed location of RV1, this was incorrect.

FMG/TPI gave approval to the RFT document prior to its release. However, the FMG/TPI representatives who signed off on the documentation did not consider themselves competent to check and approve the technical specifications. Their review and approval only focussed on the matters they had knowledge of, such as the number of dongas, location, layout of the facility, and that it had the amenities they required. Whilst they were generally aware that there were wind regions and that a set of Australian Standards dealt with this issue, none of the FMG/TPI representatives were aware of the significance of the wind region specified in the RFT or that it was incorrect.

FMG/TPI relied on Spotless to ensure the RFT was accurate and correct. Additionally, FMG/TPI representatives expected the tenderer to be vigilant in examining the RFT and its requirements, and that Spotless would be diligent in ensuring compliance with legal and regulatory requirements. This was on the basis that it would be commercially prudent for both the tenderer and Spotless to do so, and given that the building industry is highly regulated. Indeed, during the trial, Mr Smith of NT Link acknowledged that it was his duty to check whether RV1 was in Wind Region A (*Department of Consumer and Employment Protection v The Pilbara Infrastructure Pty Ltd & Anor*, 2011, [274]). Additionally, both the quote and contract between TPI and NT Link provided that it was NT Link's responsibility contractually to submit an application for the building licence with the shire. NT Link conceded that it was responsible for ensuring the application was accurate and complete.

On appeal, WorkSafe argued that it was not reasonable for FMG/TPI to make such assumptions. In particular, WorkSafe argued that it was not reasonable for FMG/TPI to assume NT Link had ultimate responsibility for the design of the dongas as that was not its role under the relevant contractual documentation (*Kirwin v The Pilbara Infrastructure Pty Ltd*, 2012, [14], [111]–[112]). A further argument of WorkSafe was that the RFT and contractual documents *required* the camp facilities to be built in accordance with the specified Wind Region A, and that NT Link had not been engaged to specify the appropriate wind region (*Kirwin v The Pilbara Infrastructure Pty Ltd*, 2012, [124]. However, these arguments were rejected by Hall J on appeal, who found that the wind region was an independent matter that the tenderer was obliged to verify, determine and comply with (*Kirwin v The Pilbara Infrastructure Pty Ltd*, 2012, [112], [115], [124]–[125], [154]).

3. *Expert contractors appeared to undertake the work competently.* The actions of the expert contractors appeared competent, and none of them, or anyone engaged by them, brought to the attention of FMG/TPI any issue concerning the design, construction or installation of the dongas. Neither did any third parties, such as the shire approving the building licence application.

At no point did anyone make known to FMG/TPI that there was an error in relation to the specified wind region. Mr Smith of NT Link attempted to check the wind region by carrying out his own measurement of the distance of RV1 from the coast. Whilst he found the distance was 97 kilometres (clearly not within Wind Region A based on the Australian Standards and Building Code of Australia), he remained silent (*Department of Consumer and Employment Protection v The Pilbara Infrastructure Pty Ltd*, 2011, [274], [329]) and took what was described by the Magistrate as a 'near enough is good enough' (*Department of Consumer and Employment Protection v The Pilbara Infrastructure Pty Ltd*, 2011, [337]) approach. That silence was relevant, because it went to the question of whether there was anything that might have alerted FMG/TPI to the issue that Wind Region A was the wrong specification (*Kirwin v The Pilbara Infrastructure Pty Ltd*, 2012, [136]–[138]).

Additionally, the shire approved all plans, diagrams and applications on the basis of the incorrect region. It was the shire's role to reject applications that did not comply with the relevant requirements. As this incorrect wind region was consistently applied throughout the entire supply process, with no issues being raised, there was nothing to suggest to

FMG/TPI that it was incorrect. On appeal, Hall J observed that this was not relied on as absolving FMG/TPI of their safety duties, but rather as a factor which was relevant in considering whether or not there were any other reasonably practicable steps for them to take, and in assessing their state of knowledge (*Kirwin v The Pilbara Infrastructure Pty Ltd*, 2012, [172]–[174]).

Further, FMG/TPI were not aware of any issues in relation to the specifications, design and or suitability of the tie downs and footings as a system of anchoring the dongas to the foundation. The tie down design and drawings were prepared by Mr Guthrie, but not checked by the engineering firm referred to in NT Link's quote, or any other engineer. It was found that the tie downs were deficient as they lacked direct horizontal constraint (*Department of Consumer and Employment Protection v The Pilbara Infrastructure Pty Ltd*, 2011, [19], [302]–[303]). However, FMG/TPI were simply not aware of any of these matters, nor did they have the expertise to identify these issues.

FMG/TPI were also not aware that there had been poor welding and other substandard installation work. Mr Lawry's supervision of the installation and building of the dongas on site did not bring anything to the attention of FMG/TPI that suggested the dongas were not suitable to be used as safe refuge. He certified the work done and progress payments were made. There was nothing to suggest there were either no or inadequate quality inspections by him or his team.

Finally, the dongas were to remain the property of NT Link's parent company, in which Mr Smith had an interest. It would be expected that NT Link would therefore be careful to ensure the buildings were designed and built to withstand a cyclone, so as to avoid damage or loss. On appeal, Hall J upheld this as a relevant factor in assessing whether FMG/TPI had satisfied its duty (*Kirwin v The Pilbara Infrastructure Pty Ltd*, 2012, [141]).

4. Lack of expertise and control by FMG/TPI. Although TPI had the contractual right to issue written directions to NT Link under the contract with NT Link, it did not have any expertise in relation to the supply of the dongas. Therefore, it could not issue any written directions about that aspect of the expert contractor's work which required expertise. However, Mr Lawry, whom FMG/TPI engaged to supervise the installation works, did give directions to NT Link to rectify deficiencies identified by Mr Lawry's team (although not all deficiencies were identified by Mr Lawry and his team).

Reliance by Laing

As already noted above, Laing was not involved in the design, construction or supply of the dongas. Laing did not have certified structural engineers on site who could assess the design specifications of the dongas, and no such assessment was ever made. Laing also did not seek any assurances that the dongas were built to withstand cyclonic winds for the relevant location.

Laing supervision and management assumed that the camp would have been built to withstand cyclones. However, they were not aware of any specific Laing inspection to determine this issue or of Laing seeking any assurance that the dongas were suitable. Their assumption was primarily based on their previous experience of living and working in the Pilbara, including that they had previously stayed in their dongas during other cyclones they had experienced. The dongas at RV1 appeared similar to others in the Pilbara, and they believed there was no cause to inspect the structural aspects of the dongas.

Additionally, Laing supervision and management were aware that WorleyParsons was the project manager/engineer for the project. They believed that the camp had been constructed for WorleyParsons, that WorleyParsons had a good reputation and WorleyParsons's quality control was good. Laing supervision and management also assumed that the construction of the camp would have been approved by the shire and that the builder would have ensured the dongas were approved correctly.

Laing's construction manager had been involved in the construction of Laing's offices at RV1, and these offices were rated to withstand Category 5 cyclones. He had seen the plans rating the offices and as far as he was aware, they were shire approved. Laing's construction manager had no reason to believe that the same process had not been followed in respect of the dongas.

Finally, under Laing's alliance with TPI, both Laing and TPI agreed to:

a. exercise due skill, care and diligence, satisfy and comply with all statutory requirements;
b. obtain all approvals, authorisations and consents that were necessary, ensure the safety and health of all persons engaged by each of them; and
c. provide and maintain a working environment where people were not exposed to hazards.

Therefore, Laing, by its representatives, assumed that the dongas had been built correctly. Further, there was nothing to suggest that the usual processes had not been followed. Indeed, shire approval suggested that all relevant regulatory requirements had been complied with to ensure the dongas were safe refuge.

No other reasonably practicable steps?

FMG/TPI and Laing had either engaged and/or relied on expert contractors and certain facts (such as shire approval) to ensure the dongas were safe refuge. Whilst the position of FMG/TPI was different to Laing, in that FMG/TPI had some level of control and knowledge of what was built and who built it, in broad terms, the courts took a similar approach in both cases. The relevant court's conclusion in both cases was that there was nothing more that either FMG, TPI or Laing could have done that was reasonably practicable (*Department of Consumer and Employment Protection v The Pilbara Infrastructure Pty Ltd*, 2011, [712]–[741]; *Kirwin v Laing O'Rourke (BMC) Pty Ltd*, 2009, pp. 90–91). The apparently qualified experts appeared to have carried out their work carefully and safely, the shire had approved the plans and issued a building licence, and there was nothing to suggest the dongas were not safe refuge.

In both cases it was held that, absent anything to put the duty holder on notice that this was not appropriate, or to indicate that the dongas were not suitable as safe refuge, it would be going beyond what was reasonably practicable to require a duty holder to engage additional experts or seek further assurances (*Laing O'Rourke (BMC) Pty Ltd v Kirwin*, 2011, [5], [67]–[69], [73]–[75]; *Kirwin v The Pilbara Infrastructure Pty Ltd*, 2012, [147], [158]–[159], [181]–[183]).

WorkSafe alleged that FMG/TPI should have engaged an engineering expert to assist it in determining the wind region specification (*Kirwin v The Pilbara Infrastructure Pty Ltd*, 2012, [143]). It argued that this was reasonably practicable and necessary to discharge the duty, in circumstances where (1) FMG/TPI (as project manager once Spotless withdrew) was responsible for the 'crucial task' of determining the correct wind specification; (2) NT Link had indicated it only took engineering advice in respect of the tie down, footings and verandah design; (3) Spotless had not utilised engineering expertise in preparing the RFT and the wind specification was not provided by an engineer; (4) NT Link had indicated in the contractual documentation that it was assuming the design specifications drafted by Spotless were correct and it was engaged on the basis of the wind region in the RFT; (5) FMG/TPI knew the location of RV1 was 90 km from the smooth coast line; and (6) there was nothing to suggest an engineer had reviewed the wind region specification (*Kirwin v The Pilbara Infrastructure Pty Ltd*, 2012, [143]–[145], [151]). However, Hall J rejected these arguments and did not agree with WorkSafe's interpretation of the facts and the contractual documents. The judge said that

> It is always possible to imagine a further step, an additional check or a second opinion that could be obtained, particularly with the benefit of hindsight. However, the context in which [FMG/TPI] operated was that they had already had the benefit of Spotless' expertise in preparing the RFT, had retained a builder with apparent expertise and

> experience and had engaged WorleyParsons for advice on engineering aspects of the infrastructure project. The question is not whether something else could conceivably be done, but whether it was reasonably practicable to expect principals in the position of the respondents to do more. (*Kirwin v The Pilbara Infrastructure Pty Ltd*, 2012, [147]).

Further, Justice Hall did not agree with WorkSafe's argument regarding the basis on which NT Link was engaged (*Kirwin v The Pilbara Infrastructure Pty Ltd*, 2012, [153]), FMG/TPI had already engaged Mr Lawry to check the quality of NT Link's work (*Kirwin v The Pilbara Infrastructure Pty Ltd*, 2012, [149]), and NT Link had represented it would obtain engineering advice and undertook to use it (*Kirwin v The Pilbara Infrastructure Pty Ltd*, 2012, [149], [156]). Justice Hall agreed with Magistrate Mignacca-Randazzo that it would have been unreasonable to expect that further experts needed to be engaged to review the work done by the experts first engaged (*Kirwin v The Pilbara Infrastructure Pty Ltd*, 2012, [85], [157]–[159]).

Similarly, in the Laing prosecution, WorkSafe submitted that Laing should have carried out its own enquiries and investigations, sought assurances, obtained engineering advice regarding the design and construction of the dongas, in assessing their suitability for cyclonic conditions (*Kirwin v Laing O'Rourke (BMC) Pty Ltd*, 2009, p. 88). It contended that these steps to assess the suitability of the dongas would not have been particularly onerous or costly, and were reasonably practicable when balanced against the severity of the potential risk from the hazard (*Kirwin v Laing O'Rourke (BMC) Pty Ltd*, 2009, p. 88). However, whilst the magistrate accepted the steps were not prohibitively expensive, he stated that all the evidence pointed away from any reason or trigger to have the dongas assessed in relation to their suitability (*Kirwin v Laing O'Rourke (BMC) Pty Ltd*, 2009, p. 89). The expert witnesses consistently stated that shire approval was commonly relied upon to assume that a building was safe refuge and built to withstand cyclonic conditions for the area (*Kirwin v Laing O'Rourke (BMC) Pty Ltd*, 2009, pp. 58–63, 83). The purpose of building approvals was said to check that the design meets applicable standards and requirements for that region. There was therefore nothing that would or should have prompted Laing to conduct further checks or seek assurances; it was not foreseeable that the dongas would be constructed to the wrong wind region, and the wrong specifications would have been approved by the relevant shire (*Kirwin v Laing O'Rourke (BMC) Pty Ltd*, 2009, pp. 83, 90–91).

On appeal, Justice Murray was persuaded by the argument that such assurances, enquiries or an engineering assessment should have been made (*Kirwin v Laing O'Rourke (BMC) Pty Ltd*, 2010, [82]–[86]). This was primarily on the basis that given the nature of the hazard and risks involved, such steps were not onerous and should have been undertaken (*Kirwin v Laing O'Rourke (BMC) Pty Ltd*, 2010, [84]–[85]). However, this was rejected by the court of appeal. The court's Chief Justice Wayne Martin noted that, taking the argument to its logical conclusion, all duty holders whose workers were placed in accommodation in cyclonic regions would be required to take similar steps, including obtaining unspecified engineering advice in relation to the accommodation, to satisfy their safety duties (*Laing O'Rourke (BMC) Pty Ltd v Kirwin*, 2011, [4]–[5]). The Chief Justice stated this was 'plainly impracticable' (*Laing O'Rourke (BMC) Pty Ltd v Kirwin*, 2011, [5]).

It is clear that each—Laing, FMG and TPI—had made certain assumptions in relation to the suitability of the dongas as safe refuge. However, Justice Hall (in the FMG/TPI appeal) stated:

> It is useful to note in this context that the fact of making assumptions is not in itself inappropriate; it is the circumstances in which any such assumptions are made that may be relevant in determining whether an employer or principal has done all that is reasonably practicable in the circumstances. It is unlikely to be enough for a person to merely assume that someone else will attend to safety requirements, but if such an assumption is based upon inquiries made, assurances given, a reasonable belief as to the skills of those responsible for construction and a reasonable belief that regulatory approval has been obtained for the buildings, it may be well-founded. (*Kirwin v The Pilbara Infrastructure Pty Ltd*, 2012, [108]).

The assumptions made by FMG/TPI were held to be reasonable, given it had retained experts throughout the process (Spotless, NT Link, Mr Lawry and Worley Parsons), and there was nothing to put them on notice to challenge the assumptions made (*Kirwin v The Pilbara Infrastructure Pty Ltd*, 2012, [105], [158]). Laing's assumptions were also found to be reasonable and based on the following: that the dongas were placed in an area known to be subject to cyclones, the alliance responsible for producing the specifications was contractually committed to safety, the shire had a role to play in the enforcement of building requirements and was familiar with the requirements applicable to the shire region, and previous experience pointed to dongas being used as shelter during a cyclone (*Laing O'Rourke (BMC) Pty Ltd v Kirwin*, 2011, [46]–[52], [67]–[69]).

The FMG/TPI prosecution disclosed that NT Link had, like FMG/TPI and Laing, also made a number of assumptions, and had failed to check or verify matters that FMG and TPI were relying on NT to do, and had failed to engage engineering experts to assist it in its work. In particular, NT Link assumed that the Shire of East Pilbara would, for itself, investigate whether RV1 was indeed in wind region A (*Department of Consumer and Employment Protection v The Pilbara Infrastructure Pty Ltd*, 2011, [315]). NT Link also relied on the RFT issued by Spotless and assumed Spotless would have done an assessment and ensured the information in the RFT was correct (*Kirwin v The Pilbara Infrastructure Pty Ltd*, 2012, [145]). Justice Hall stated that the failures by NT Link did not necessarily lead to a conclusion that FMG/TPI failed in their duty (*Kirwin v The Pilbara Infrastructure Pty Ltd*, 2012, [159]). Justice Hall found that FMG/TPI had not neglected to consider the issues of engineering and design in regard to cyclone safety (*Kirwin v The Pilbara Infrastructure Pty Ltd*, 2012, [150]). Rather, they had sought to engage experts to assist them with these matters (*Kirwin v The Pilbara Infrastructure Pty Ltd*, 2012, [150], [159]). This raises the question of whether parties should take precautionary steps on the basis that expert contractors may also make errors, which may potentially impact on the safety of workers, as occurred in the Cyclone George disaster.

In the Laing appeal before the court of appeal, WorkSafe argued that the requirement that duty holders take an active, imaginative and flexible approach to potential dangers, with the knowledge that human frailty is an ever-present reality, meant that it was not sufficient for employers to assume that expert contractors or other third parties would always act competently (*Laing O'Rourke (BMC) Pty Ltd v Kirwin*, 2011, [42]). Rather, duty holders needed to be alert to the possibility that contractors and shires, not just employees, may act in a way, or make errors, that impact on safety (*Laing O'Rourke (BMC) Pty Ltd v Kirwin*, 2011, [42]). However, Murphy JA was not prepared to accept that the fact that the shire and the expert builder both made errors was foreseeable in the circumstances (*Laing O'Rourke (BMC) Pty Ltd v Kirwin*, 2011, [67]–[69]). His Honour emphasised there was nothing to suggest to Laing that the dongas were unsuitable refuge, or that a reasonable employer would have appreciated or foreseen that the accommodation posed a risk in the event of a cyclone (*Laing O'Rourke (BMC) Pty Ltd v Kirwin*, 2011, [68]). Further there was no evidence to suggest that (1) a structural engineer would have discovered the errors, or (2) that if any enquiries had been made, Laing would have been told anything but that the dongas had been constructed in accordance with plans and specifications approved by the shire and in accordance with building standards, or (3) that the local authority would have provided any other information, or that any response would have revealed information to challenge the assumption that the dongas were safe refuge (*Laing O'Rourke (BMC) Pty Ltd v Kirwin*, 2011, [73]–[75]). Justice of Appeal Murphy in the Laing appeal and Justice Hall in the FMG/TPI appeal both considered that the errors in the assumptions only emerged with the benefit of hindsight (*Laing O'Rourke (BMC) Pty Ltd v Kirwin*, 2011, [70]; *Kirwin v The Pilbara Infrastructure Pty Ltd*, 2012, [181]–[182]).

Drawing the threads together

The Cyclone George cases demonstrate that the measures which a principal must take to discharge its safety duties will be a matter of fact and vary from case to case, depending on what is reasonably practicable in the circumstances.

What we see emerging from these cases is the balancing exercise that lies at the heart of safety law in Australia, between ensuring the safety of those at the workplace by eliminating risks to safety on the one hand, and recognising the practical reality that it may be commercially unviable for a duty holder to take certain steps to eliminate or reduce hazards to safety. Duty holders must determine what steps are available to mitigate risks to safety, and whether these steps are reasonably practicable. This calls for a risk assessment and value judgment. The evaluation of what is reasonably practicable is made in a particular context or operating environment, and this is relevant to the assessment of what is reasonable (*Kirwin v The Pilbara Infrastructure Pty Ltd*, 2012, [181]–[182]). This context may include that multiple experts have been engaged for various aspects of project work, the detail in what has been contractually agreed regarding parties' responsibilities, and the regulatory environment (for example, one aimed at ensuring buildings constructed in cyclone prone areas are constructed to a standard commensurate with the applicable degree of risk). There is therefore a tension between ensuring workers are safe and what is legally required of duty holders, which is necessarily a more limited duty, to take account of commercial realities, and be achievable in light of the consequences of non-compliance.

In the case of using expert contractors where a party does not have the relevant specialist expertise, the balancing exercise may involve a consideration of whether it is reasonably practicable for a party to supervise or check, or engage an expert to supervise or check, the work of the first-expert contractor engaged, and perhaps another expert to check that supervising expert. Plainly, the risk that contractors, including expert contractors, may make mistakes is ever present. However, the courts in the Cyclone George cases were reluctant to find that it was reasonably practicable to seek additional expert advice to ensure that any mistakes or inadvertent actions of an expert contractor were captured and rectified, unless there was something to trigger a suspicion about, or put the duty holder on notice of, some risk with the whole or an aspect of the expert contractor's work.

This approach is cognisant of the limited knowledge a duty holder has to recognise or be aware of potential safety issues or risks associated with the expert contractor's work (which by its nature is specialised), and the duty holder's limited control to issue safety directions, as compared with its own employees and systems of work. An approach that would require duty holders to always assume expert contractors may fail would be too broad.

Towards a collaborative approach

The above discussion illustrates that one of the key failings in the Cyclone George disaster was that each party assumed that someone else would take responsibility for ensuring that the relevant specifications were correct. Whilst Laing, FMG and TPI were found to have acted reasonably in their reliance and assumptions, the case is a good example of the gaps that can occur in complex transactions involving multiple parties.

Whilst sometimes these gaps only emerge through the benefit of hindsight, it is argued that there are at least two measures which parties can implement to minimise the risk of these kinds of gaps occurring. The first is for parties to adopt a comprehensive contractor safety management approach, starting from the beginning of the tender and engagement process, and right through to post-completion. Such an approach has been advocated by a number of different commentators, and is adopted across most major projects (Inns 'Contractors' in Dunn and Chennell, 2012; Tooma, 2011; McCartney, 2012).

Second, a collaborative approach with clear communication regarding responsibilities and reliance assists in ensuring each party is aware of where it is placing reliance on a party for a particular safety matter, or where another party is placing reliance on it. In this regard, it is important to note that as part of the move towards a harmonised, national workplace health and safety (WHS) system, the model workplace health and safety legislation has introduced a new, express obligation on parties to consult, cooperate and coordinate activities (Workplace Health and Safety Act 2011 (Cth), s 46). The duty to consult with other concurrent duty holders places a positive obligation to engage in proactive dialogue in relation to safety and health issues. This would include a dialogue between parties relying on expert contractors and expert contractors. That is not to say that these parties are not already engaging in a dialogue about safety matters. But an express obligation in the WHS legislation at least brings this issue to the forefront of parties' minds.

It should be observed that it may be difficult to measure whether a positive obligation to consult will have any tangible effects in relation to safety outcomes or processes. Further, this duty is also limited by the qualifying words 'as far as is reasonably necessary' (Workplace Health and Safety Act 2011 (Cth), s 46). This may reduce the duty's effectiveness. However, by expressly requiring a dialogue to occur between concurrent duty holders, the legislation at least encourages parties to consider and discuss the issues raised in this chapter, and it is hoped that this may assist in identifying issues or gaps at an early stage, allowing them to be rectified.

References

Baiada Poultry Pty Ltd v The Queen (2012) HCA 14.

Campbell, K. (2010, May 25). Charges Dropped over Cyclone Death Dongas. The West Australian. Retrieved from http://au.news.yahoo.com/thewest/a/-/newshome/7287959/charges-dropped-over-cyclone-death-dongas/.

Department of Consumer and Employment Protection v The Pilbara Infrastructure Pty Ltd & Anor (Magistrate's Court of Western Australia, Magistrate Mignacca-Randazzo, 18 February 2011).

Hamersley Iron Pty Ltd v Robertson (unreported, Sup Ct of WA, Steytler J, Library No 980573, 2 October 1998).

Holmes v R E Spence & Co Pty Ltd (1993) 5 VIR 119.

Inns, M. 'Contractors' in Dunn, C. and Chennell, S. (2012). *Australian Master Work Health and Safety Guide*. Sydney: CCH Australia Ltd.

Kirwin v Laing O'Rourke (BMC) Pty Ltd (Magistrate's Court of Western Australia, Magistrate Malone, 30 October 2009).

Kirwin v Laing O'Rourke (BMC) Pty Ltd {2010) WASC 194.

Kirwin v The Pilbara Infrastructure Pty Ltd (2012) WASC 99.

Kondis v State Transport Authority (1984) 154 CLR 672.

Laing O'Rourke (BMC) Pty Ltd v Kirwin (2011) WASCA 117.

McCartney, S. (2012). Can you trust your independent contractor with WHS obligations? Australian Occupational Health and Safety Tracker. CCH Australia Ltd.

R v ACR Roofing Pty Ltd (2004) 11 VR 187.

Tooma, M. (2011). *Safety, Security, Health and Environment Law* (2nd ed.). Sydney: The Federation Press.

chapter eight

Contractor safety, health and environmental regulations in the MENA region

Elias M. Choueiri, PhD*
President
Lebanese Association for Public Safety

Contents

* High-ranking executive officer in both public and private institutions in Lebanon. Coordinator/professor at Sagesse University (ULS) and board member, World Safety Organization, USA.

Introduction

The MENA infrastructure and construction market is amongst the world's most attractive, given its sheer size. Forecasting figures predict a total of $4.3 trillion will be invested in construction projects across the MENA region by 2020, representing an increase of almost 80% from today's spending ('Saudi Arabia to Lead MENA Construction Boom', www.constructionweekonline.com, 2012).

The region is expected to account for 12% of the global emerging markets and 4.4% of the world construction markets within the next decade with Saudi Arabia expected to continue leading the way. Although MENA contract awards have declined by 41% so far in 2012 from a year earlier, the main reasons can be primarily attributed to delays in awarding petrochemicals projects in Egypt and in awarding construction and infrastructure contracts in the UAE, Kuwait and Iraq. The construction and infrastructure sub-sectors in Saudi Arabia, however, remain strong, growing by 177% over the same period, and currently accounting for 46% of the 2012–2013 MENA project pipeline, totalling $448 billion. With its young and expanding population, Saudi Arabia should remain the most buoyant market, in line with its overall economic development plan. Furthermore, the recent approval of the mortgage law should help to drive growth in residential construction in response to the current housing shortage (Saudi Arabia to Lead MENA Construction Boom, 2012).

By all means, construction contracting in MENA could simply be described as regional departures from international practices (Dimitracopoulos, 2008). In the United Arab Emirates (UAE), for instance, contractors are lulled into a false sense of security, thinking that they have budgeted for all delay-related liquidated damages that may be imposed upon them; the contract specifies the daily amount and, in any event, there is a cap on it of usually no more than 10% of the contract value. The contractor would, of course, try to avoid any delays—or at least avoid responsibility for them—but as long as they have factored that 10% into *their figures, the worst-case scenario, they feel, has been forecasted and budgeted for.*

- *This perception would probably be right in MENA; contracts do almost always include a liquidated damages clause, and the parties agree that a predetermined daily amount would be subtracted from the contractor's monthly invoices if, in the opinion of the employer, the contractor has been late (subject always to a maximum total cap of a percentage of the contract—usually 10%, or 5% or less in very large contract values).*
- *There are, however, some variations to this arrangement in MENA, relating mainly to the desire some employers have to appropriate that daily amount whether they have actually been affected by the delay or not. It is common to see attempts seeking to set out the parties' true intentions by using words along the lines of '… without the employer having to prove actual loss' or '… regardless of whether the employer has incurred any loss'.*
- *In short, parties sign off on the deal, and in the mind of the contractor, the good news would hopefully be that it will never have to pay more than 10% of the contract value, no matter how late the project is. In the mind of the employer, the good news seems to be that, if there is any delay, it can arguably allege it is the fault of the contractor, and then enjoy an automatic discount on the original contract price.*

Having said all that, MENA law, like most civil law–based codified legal systems, includes provisions that could surprise the unwary contractor or employer given the right set of circumstances.

Dimitracopoulos notes the following:

- *For a start, in MENA law, the concept is generally upheld that damages actually incurred by an employer cannot be lower in quantum than the liquidated damages deducted. Allowing the employer to deduct a predetermined amount in the event of delay, regardless of whether such delay has actually affected the employer proportionately, would arguably be tantamount to undue enrichment. As such, it would be subject to scrutiny by the contractor's lawyers who would be asked to put the employer to strict proof of not only who the culprit for the delay was but also of whether that delay actually made a financial and measurable difference to the employer.*
- *In practice, the employer has the initial privilege of simply activating the liquidated damages clause and reducing the amount it considers payable under the contractor's invoices by the agreed daily amount. This places the contractor in a defensive position as it would have to go through the whole process of filing its claim with the engineer and—if the amount justifies it—referring the issue to arbitration.*

- *In that scenario, the employer may find that, although it has stated that deduction would take place regardless of any loss incurred, it could be asked to prove more than whether the delay was attributable to the contractor; it may have to discharge the burden of proving that this delay has caused it quantifiable actual loss and, if not expressly excluded, consequential loss. In most cases this would be a difficult hurdle to overcome for the employer, as any evidence adduced (pointing to loss of income, for example) would in all likelihood be circumstantial. This could mean that the financial pressure initially applied to the contractor by the employer's application of liquidated damages might in the end prove to be a pyrrhic victory. Any amounts withheld and not applied to provable loss may have to be returned to the contractor.*
- *There are arguments expressed in the UAE and elsewhere in the MENA region that the onus of proof is not actually on the employer (that it has incurred loss) but on the contractor (that the employer has not). Whatever the case may be, the message to be borne in mind for the employer is that a liquidated damages clause is not an irrevocable right for deductions to be made. It would ultimately need to prove that it has actually incurred a loss along the lines the parties had initially contemplated would actually be incurred. Alternatively, it may have to counter evidence adduced by the contractor that no such loss has actually been incurred.*
- *The provision of liquidated damages is not considered in MENA as an unassailable predetermination of actual losses to be incurred, and it is always open to the parties to revisit the issue and adduce evidence that would prove presence or absence of an alleged loss.*

It is important to note here that all is not easy for the contraction either. In this respect, Dimitracopoulos highlights the following:

- *The maximum exposure that it had initially factored in could, under MENA law, be exceeded if the delay is substantial and if it resulted in losses being incurred by the employer that are not only quantifiable but are also well in excess of what were originally envisaged as likely.*
- *Prudent and locally seasoned international contractors often question in advance the validity of their anticipated contractual arrangements. Particularly those that are more worldly wise and have experienced different treatments of liquidated damages clauses around the globe, often ask whether they can indeed rest assured that when the contract says 10% maximum for delay damages, this is always going to be the case.*
- *Whilst this judicial discretion (available also to arbitrators) is rarely applied in practice, the legal provision opening the door for breaking limits of liability is available in MENA. Like any other mandatory law provisions, it will apply to all construction contracts. No matter how 'standard', they are all subject to the overriding laws of a given jurisdiction.*

- *At least with regard to liquidated damages, which can mean different things in various countries, it may be wise for both the employer and the contractor to be clear on their intentions. They could set out in the contract a list of the specific and foreseeable instances that may lead to a tangible loss for the employer and which the liquidated damages clause is in each case meant to address.*

On the backdrop of an unprecedented construction boom witnessed in MENA, it is fair to say that the landscape of construction law and practice is in the process of being shaped by parties moulding their intentions into better-thought-out contracts. It is encouraging to see contractors, as well as architects, throughout the MENA region become increasingly aware of the significant risks in treating international contracts as a universal and infallible text transcending cultures, laws and practices; see, for example, the Appendix 'Contractor Safety, Health and Environmental Regulations' as adopted by the Council for Developement and Recontruction, Lebanon.

Background

Occupational safety and health (OSH) in MENA

In view of the drastic developments that are taking place in MENA, responsible companies in the region* (see Figure 8.1), have realised that occupational safety and health (OSH) is an integral part of business performance, and if they are not organised for OSH, they will not be profitable or sustainable. Today, there are numerous grounds to indicate that OSH is becoming considerably significant across all industry sectors in

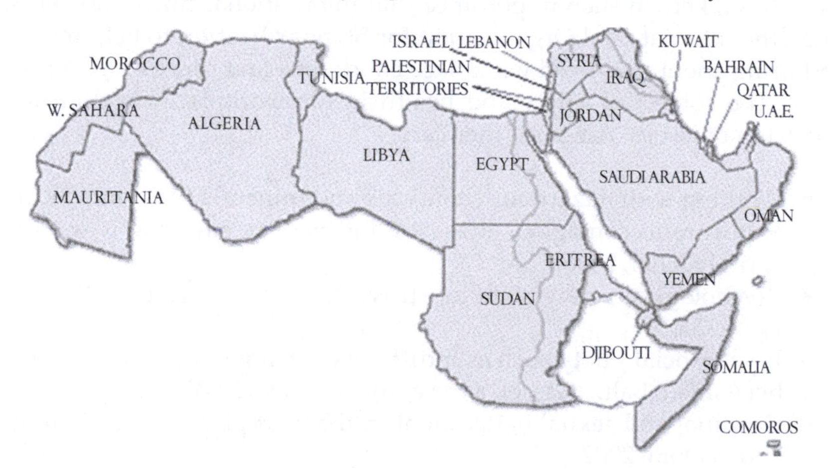

Figure 8.1 MENA countries.

* According to the World Bank, MENA has a population of 355 million, with 85% living in middle-income countries, 8% in high-income countries and 7% in low-income countries.

the MENA region. A good number of MENA organizations, especially the large ones, have commenced to efficiently implement OSH management systems in order to protect the workforce and promote a healthy working environment. They are striving to build a safety culture in a changing environment, and inculcate employees with responsibility towards their own safety. The importance of committed leadership in driving a safety culture and the role of government in influencing this are needs-of-the-hour issues that a good number of conferences in MENA have been trying to bring to light ('Imperative Need for Occupational Health and Safety in Middle East', Middle East OHS Strategy Summit, 2012).

It should be noted that there are many challenges and institutional gaps impeding effective prevention of occupational accidents and diseases in MENA. Amongst these are the lack of resources devoted to OSH, including the provision of services and awareness raising, the low rates of ratifications of international labour conventions, a lack of national capacity in training, retraining and further developing the skills of OSH specialists, weaknesses in the labour inspection components and the absence of clear national plans. Although many MENA countries have comprehensive laws concerning basic working conditions, including occupational health and safety standards and minimum wages, their enforcement is weak and the national OSH systems are insufficiently developed (Overview of the Occupational Safety and Health Situation In the Arab Region, ILO, 2007).

Safety in the workplace

Because of globalised economic trends, the subject of safety in the workplace has taken on such importance that international conventions instituted the International Organization for Standardization to help regulate and bring about improved workplace conditions and services (Zwetsloot, 2003). The subject of safety and health in the workplace covers a wide spectrum of issues. Amongst them are:

- Working with hazardous chemicals and minerals (Armour, 2003).
- Exposure to contagious diseases and passive smoking (Gwandure and Thatcher, 2006).
- Psychological safety such as stress, fears and attitudes (Baer and Frese, 2003).
- Psychosocial safety such as indifference, xenophobia and homophobia (Gillen, Baltz, Gassel, Kirsch and Vaccaro, 2002).
- Criminal and sexual harassment in the workplace (Hatch-Maillette and Scalora, 2002).
- Working within harmful workplace emissions (Profumo, Spini, Cucca and Presavento, 2003).
- Manufactured and manufacturing of harmful substances and innovations (Valent, McGwin, Bovenzi and Barbone, 2002).

- Harmful infrastructural constructions such as unsafe stairways, unsafely built structures and slippery floors (Mehta and Burrows, 2001).
- Terroristic intrusions and massacres in the workplace (Miller, 2001).
- Safety precautions, safety communication measures and personal protection equipment (Henshaw, Gaffney, Madl and Paustenback, 2007).

Rules on working time and rest periods in MENA

In MENA countries, working time is determined by legislation, although in some cases more favourable terms can be laid down by collective agreement in certain sectors. The law does not elaborate on the concept of working time; rather, it sets out maximum weekly working hours, which vary from country to country. The number of hours per week is frequently 48, but it is lower in certain countries, for example, the Palestinian Territories, where weekly working time totals 45 hours, and in Algeria where the normal working week is 40 hours (Mzid, 2004).

Similarly, the length of the working day is fixed by law. In this respect, however, legislation in MENA countries differs considerably. In a nutshell, actual working time is usually set at eight hours per day. But this is very variable, on account of certain special derogations allowing the working day to be lengthened beyond that limit for certain types of activity, or—with authorization from the labour inspectorate—through the use of overtime. Overtime is paid at a higher rate, which is established by law and varies from one country to another.

The different MENA countries also have specific provisions concerning certain groups of workers, most notably women and young people. For instance, women and children are often forbidden to work at night. Yet, this prohibition, far from being absolute, is tempered by a myriad of legally regulated derogations.

In terms of rest periods, legislations in all MENA countries sanction:

- The right to weekly rest of 24 consecutive hours.
- The right to an annual leave, although there are considerable differences in its duration from one country to another. It is important to note here that, in order to protect the right to annual leave, the law sometimes rules null and void any agreement whereby the employee waives this right, even in return for financial compensation. However, for want of effective enforcement, it can happen that employees do 'waive' their right to annual leave and so continue working throughout the year without any opportunity to rest.
- The right to maternity leave, granted to women on the birth of a child. During this leave, the duration varying from country to country, the employment contract is suspended. In some countries the woman is

entitled to continue receiving her wages from the employer; in others, a compensatory allowance is paid by the social security fund. Legislation in all the countries stipulates that a woman's time-off for reasons of childbirth should not lead to termination of her employment contract, but dismissal is often sanctioned by the payment of damages rather than by being declared null and void. Furthermore, only in Morocco are there penal sanctions to protect the job of a woman on maternity leave.

- The right to time-off with pay on public holidays. These are rest days offered in commemoration of certain national or religious events.
- The right to special leaves that may be granted for a variety of reasons: family events, fulfilment of a legal or religious obligation, participation in trade union activity, etc.

It should be noted that labour legislation throughout MENA provides protection for work-related injury and disability as part of social security provisions. However, the number of workers with access to these benefits is small, since large segments of the labour force work in the informal sector without access to social security provisions. Today in MENA, and despite a comprehensive legislative framework, the majority of workers remain unprotected to the risks of work-related injury and disability ('A Note on Disability Issues in the Middle East and North Africa,' The World Bank, 2005).

Regulatory texts concerning safety and health in the workplace in MENA

It would not be overstating the case to say that, in the area of health and safety, in particular, the legislation in MENA countries is fairly ineffectual. Despite a whole host of detailed regulatory texts concerning health and safety in the workplace, in actual fact the prevention of occupational risk is generally anything but adequate (Overview of the Occupational Safety and Health Situation In the Arab Region, 2007).

From the authors' experience in safety work, training and education, there has been minimal interest in occupational safety, except for specific companies or large companies. Small and medium-sized companies in MENA do not show the slightest interest in safety, except when they are forced to for one reason or another. The community as a whole can be regarded as ignorant when it comes to safety issues. Many MENA countries have not yet reached a safety culture, but only know the basics, maybe because of lack of safety professionals in the workplace safety area.

By all means, some MENA employers assume little responsibility for the protection of workers' health and safety. In fact, some employers do not even know that they have the moral and often legal responsibility to protect workers.

Updating national workplace safety legislations in MENA

In relation to updating National legislations, all MENA countries have amended their legislations and labour codes at various levels to match international standards and improve the working conditions of the labour forces. The updates are mainly related to the labour laws in the public sector, the social security codes, the rights of the handicapped, the agricultural laws, child labour, inspection, and other OSH issues that had been initially absent from local legislations (Overview of the Occupational Safety and Health Situation In the Arab Region, ILO, 2007).

Nowadays, MENA countries-based organizations are being forced to review their health and safety policies in view of the impending changes in healthcare legislation, which will result in companies becoming responsible for their employees' healthcare costs. This significant development means that corporations will have to adopt a responsible and proactive approach to employee health. Additional developments are likely to be brought about by the establishment of Occupational and Environmental Medicine Groups, which seek to develop and standardize occupational and environmental policies throughout the MENA region (Overview of the Occupational Safety and Health Situation in the Arab Region, ILO, 2007).

Authorities responsible for drafting OSH laws in MENA

The Ministry of Labour (MOL) is the key authority in charge of OSH legislation in MENA countries (Overview of the Occupational Safety and Health Situation in the Arab Region, ILO, 2007; Occupational Safety and Health, ILO, 2011). In some countries, such as in Bahrain, Morocco, Kingdom of Saudi Arabia (KSA), and Syria, the Ministry of Health is also involved; in the UAE and Yemen, the Ministry of Justice is also involved.

Besides ministries, there are institutions in some countries that are also involved in OSH legislation, such as in Syria, KSA and others. Workers' and employers' unions are also involved in OSH legislation in countries such as Oman, Palestine, UAE and Yemen.

Enforcement and implementation of OSH laws

MOL is also the key authority involved in the implementation of OSH laws and regulations in MENA countries; in some countries, the ministry of health is also involved. In addition, there are certain institutions in some countries that take part in law enforcement, such as civil defence, Social Security institutes, labour inspectorates, municipalities, and OSH bureaus (Overview of the Occupational Safety and Health Situation in the Arab Region, ILO, 2007; Occupational Safety and Health, ILO, 2011).

OSH legislation coverage of the workforce

The percentage of OSH legislation coverage of the economically active population varies from one country to another; yet, in most countries, the legislation states that all workers should be covered. However, implementation would be exclusive to certain economic sectors in some countries, or certain cities and districts in others. For instance, the informal sector is not covered in several countries such as Algeria, Kuwait, Syria and Palestine. In other countries, family businesses, the military and the public sector, and the agricultural sectors are not covered. In Lebanon, for instance, coverage is mainly in the vicinity of the capital and the major cities, leaving the rest of the areas insufficiently covered (Overview of the Occupational Safety and Health Situation in the Arab Region, ILO, 2007; Occupational Safety and Health, ILO, 2011).

OSH inhibitors in MENA

Generally speaking, management in most MENA countries lacks awareness of occupational safety and health, its core matter, its relevance and significance in the workplace. By all means, there is no perception of risk or risk assessment, as alluded to in the following (Azzi, 2009):

- Workers who have very low education are not worth training on OSH and will not understand it.
- Even if management were to enforce safety and health measures, there is high worker non-compliance.
- A direct link between a safe and healthy environment and productivity is just not clear; the benefits to management do not seem to be tangible.
- Lower management generally complains that there is no higher management commitment in terms of enforcement and follow-ups.
- Money is worth worrying about, rather than to worry about work conditions, which are regarded as secondary, as shown in Figures 8.2 and 8.3.
- Economic and political instability in the region bring the priorities down to the basic needs of survival, in light of which OSH is considered a luxury.
- Ignorance leads to the lax attitude that management has towards occupational safety and health; this in turn results in poor safety and health practices.

Occupational accidents

The following statistics are obtained from the report 'Key Global Statistics on Safety at Work' (Occupational Safety and Health, ILO, 2011):

- Each day, an average of 6,000 people die as a result of work-related accidents or diseases, totalling more than 2.2 million work-related deaths a year. Of these, about 350,000 deaths are from

Figure 8.2 A very chaotic construction site.

Figure 8.3 Unsafe construction site, in addition to non-safety precautions by construction workers.

workplace accidents and more than 1.7 million are from work-related diseases. In addition, commuting accidents increase the burden with another 158,000 fatal accidents.

- Each year, workers suffer approximately 270 million occupational accidents that lead to absences from work for 3 days or more, and fall victim to some 160 million incidents of work-related disease.
- For every case of death, 500 to 2,000 work-related injuries take place.
- Approximately 4% of the world's gross domestic product is lost with the cost of injury, death and disease through absence from work, sickness treatment and disability and survivor benefits due to accidents and poor working conditions.

- Hazardous substances kill about 438,000 workers annually, and 10% of all skin cancers are estimated to be attributable to workplace exposure to hazardous substances.
- Asbestos alone causes about 100,000 deaths every year and the figure is rising annually. Although global production of asbestos has fallen since the 1970s, increasing numbers of workers in the United States of America, Canada, United Kingdom, Germany, Australia and other industrialised countries are now dying from past exposure to asbestos dust.
- Silicosis—a fatal lung disease caused by exposure to silica dust—still affects tens of millions of workers around the world. In Latin America, 37% of miners have some degree of the disease, rising to 50% amongst miners aged over 50. In India, over 50% of slate pencil workers and 36% of stonecutters have silicosis.

Statistics on safety at work in MENA

Most of the MENA countries do not have regular statistics (Overview of the Occupational Safety and Health Situation in the Arab Region, 2007; Occupational Safety and Health, ILO, 2011). Whilst some are sporadic, others are annual and often inaccurate. The mechanism of reporting occupational accidents in MENA varies from one country to another. In Bahrain, Jordan, Lebanon, Sudan and Yemen, for example, an action notification form is to be filled out and handed to the authorities within 24 hours of accident occurrence. On the other hand, enterprises in Egypt must report occupational accidents to the appropriate authorities every 6 months.

Generally speaking, data provided by MENA countries on occupational accidents lack precision. For instance, whilst countries like Tunisia, Syria, Sudan, Kuwait, Qatar and Oman indicate that 90% to 100% of occupational accidents are reported, other countries in the region, such as Morocco, Saudi Arabia, Egypt, UAE and Yemen, report quite low casualty figures.

In countries where occupational accident reporting takes place, the exclusion of certain sectors is common. For example, family businesses, governmental, public, agricultural and informal sectors (i.e., micro-enterprises, domestic services, and self-employed workers) are not covered in UAE, Syria, Jordan, Kuwait, Lebanon and Morocco.

According to the ILO (Overview of the Occupational Safety and Health Situation in the Arab Region, ILO, 2007; Occupational Safety and Health, ILO, 2011), close to 19,000 work-related fatalities occur in Middle Eastern ME countries alone. The construction industry of the MENA region accounts for a large number of occupational accidents. The most common types of accidents observed in the industry, as in other countries (Iunes, n.d.), are caused by:

Figure 8.4 Unsafe manoeuvres and non-safety precautions by construction workers on a ladder.

- Falls from ladders, lifts and scaffolds (see, for example, Figures 8.3 and 8.4). Falls consistently account for the greatest number of fatalities in the construction industry. These types of accidents often involve a number of factors, including unstable working surfaces, the misuse of fall protection equipment, workers slipping or being struck by a falling object. The use of guardrails, fall arrest systems, safety nets and covers can prevent many such deaths and injuries.
- The use of defective or negligently operated cranes, hoists and derricks. Many of these accidents are preventable and are usually caused by poor safety procedures and negligence;
- The use of dangerous equipment, tools and machines. Moving machine parts have the potential for causing severe workplace injuries, such as crushed fingers or hands, amputations, and burns and blindness, amongst others. These injuries can be prevented with the use of equipment with appropriate design and protective features, along with training in safe operation.
- The use of explosive, corrosive and poisonous gas. Many construction projects require the use of compressed gases, which may be combustible, explosive, corrosive, poisonous, or pose some combination of hazards. The safe design, installation, operation and maintenance in accordance with the appropriate codes and standards are essential to worker safety and health.

Whilst these risks are common to construction in all MENA countries, they are exacerbated by the region's climate (heat and humidity) and particularly by the lack of adequate protection and training.

Table 8.1 Fall deaths in MENA countries—2010

Country	Fall deaths	Fall deaths per 100,000 inhabitants
Algeria	737	3.67
Bahrain	n.a.	n.a.
Comoros	11	2.97
Djibouti	37	5.19
Egypt	1,376	1.80
Eritrea	79	3.72
Iraq	967	5.30
Israel	103	1.18
Jordan	198	4.64
Kuwait	55	1.91
Lebanon	140	3.75
Libya	159	3.51
Mauritania	67	3.88
Morocco	637	2.54
Oman	84	3.27
Qatar	34	3.17
Saudi Arabia	1,391	6.58
Somalia	568	6.69
Sudan	1,563	4.64
Syria	168	1.18
Tunisia	214	2.37
United Arab Emirates	84	1.69
Yemen	820	5.26
Total	**9,492**	

Available statistics on fatalities from falls in the construction industry in MENA countries are shown in Table 8.1, along with fatalities per 100,000 inhabitants, as derived from World Health Rankings (2011), LeDuc Media, USA.

The fatality rate figures (fall deaths per 100,000 inhabitants) shown in this table are presented in Figure 8.5, from highest to lowest rates. As can be seen from the figure, Somalia has the highest fatality rate of 6.7 fatalities per 100,000 inhabitants, and Israel has the lowest fatality rate of 1.2 fatalities per 100,000 inhabitants.

It should be noted that a good number of people who work in the construction industry in the MENA region are migrant workers. Many of the estimated 20 million migrant workers in MENA are from poor countries whose leaders have long failed to put in place mechanisms to protect their nationals from abuse, inhumane working conditions, and trafficking and to provide a means for repatriation during times of crisis (Russeau, 2011).

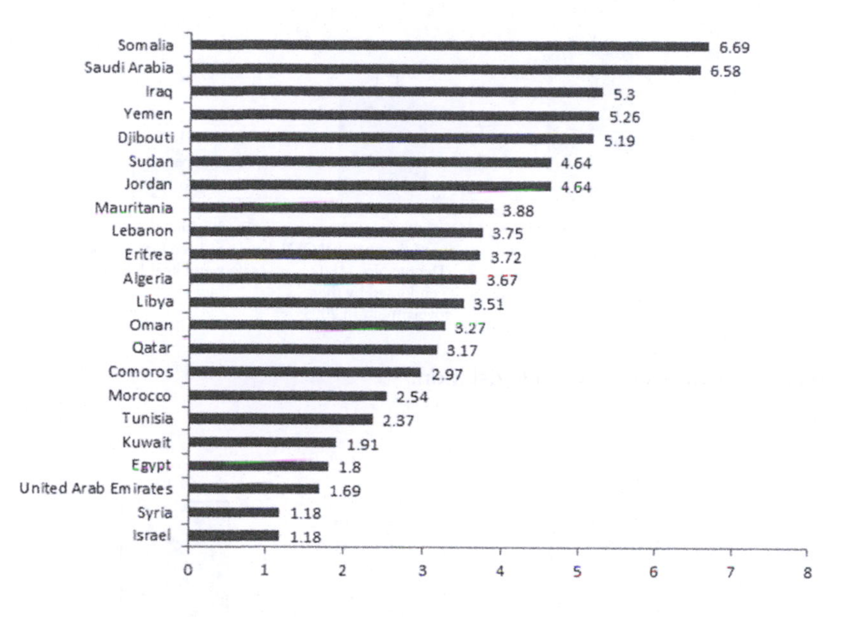

Figure 8.5 Fall deaths per 100,000 Inhabitants in MENA countries—2010.

Generally speaking, migrant construction workers in MENA countries are often subjected to overcrowding and unhygienic living conditions, and corrupt recruitment brokers may take a cut from their wages. They often work in unsafe conditions and extremely high temperatures (http://www.migrant-rights.org/faq/).

Perceptions concerning occupational safety and health

A study concerning 'Challenges in Incorporating Safety and Health into National Plans: Between Policy and Practice in Lebanon' (Overview of the Occupational Safety and Health Situation in the Arab Region, ILO, 2007), the results of which could simply be extrapolated to other countries in the MENA region, considering that all the countries in the region share similar characteristics, yielded the following results:

Awareness of OSH legislation (see Figure 8.6):

- More than 90% of heads of enterprises stated that they were not aware that ILO has conventions on OSH.
- More than 85% of heads of enterprises were not aware of the national OSH Decree No. 11802* endorsed by the government in 2005.

* Decree No. 11802: Regulates occupational prevention, safety and health in all enterprises subject to the code of labour. It entails five chapters: Chapter 1 deals with 'Prevention and Safety'; Chapter 2 deals with 'Health'; Chapter 3 deals with 'Safe Use of Chemicals at Work'; Chapter 4 deals with 'Prevention from the Dangers of Working with Benzene'; and Chapter 5 deals with 'General Provisions'.

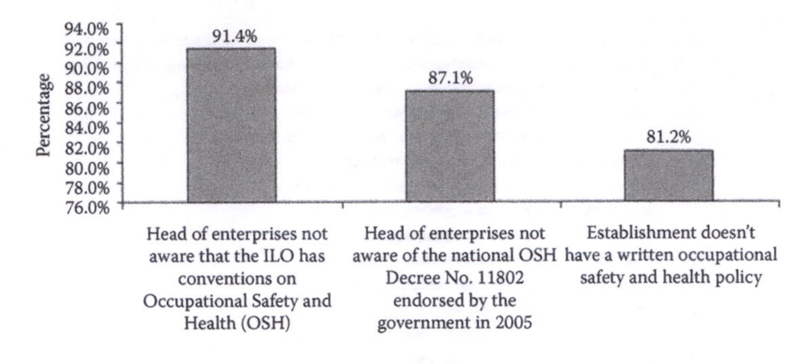

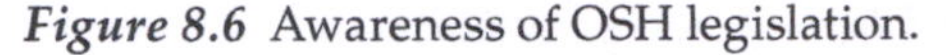

Figure 8.6 Awareness of OSH legislation.

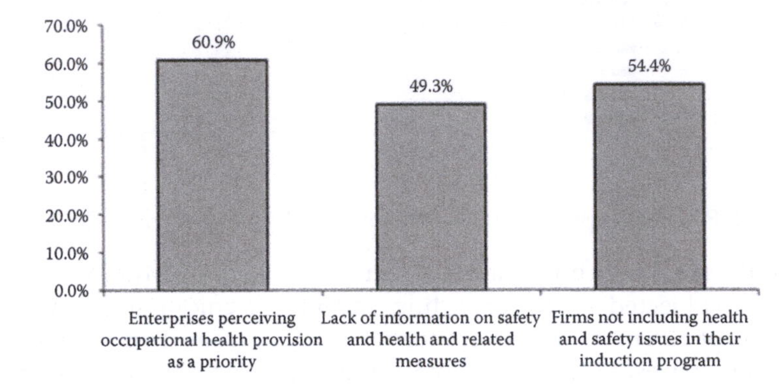

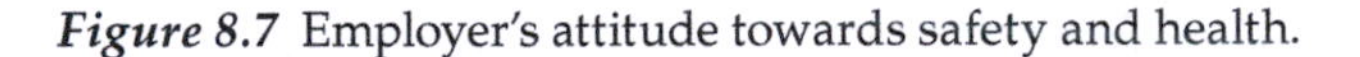

Figure 8.7 Employer's attitude towards safety and health.

- More than 80% of establishments did not have a written OSH policy.

Employers' attitude towards occupational safety and health (see Figure 8.7):

- More than 60% of enterprises perceived occupational health provision as a priority.
- Close to 50% of employers lacked information on safety, health and related measures.
- Close to 55% of firms did not include health and safety issues in their induction programs.

Motivation for providing OSH at the workplace (see Figure 8.8):

- Close to 70% of respondents indicated that the safety of employees means higher productivity.

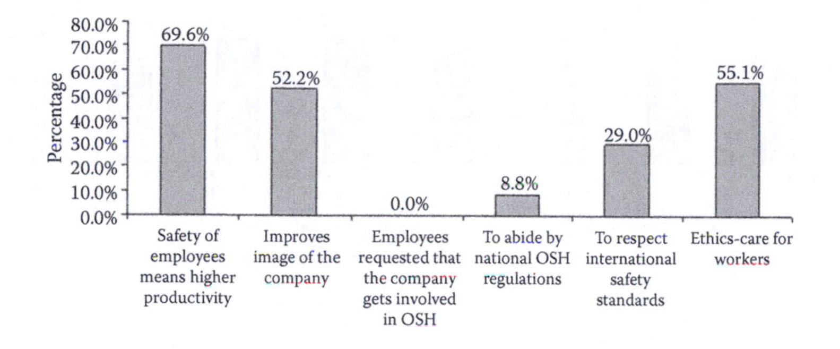

Figure 8.8 Motivation for providing OSH at the workplace.

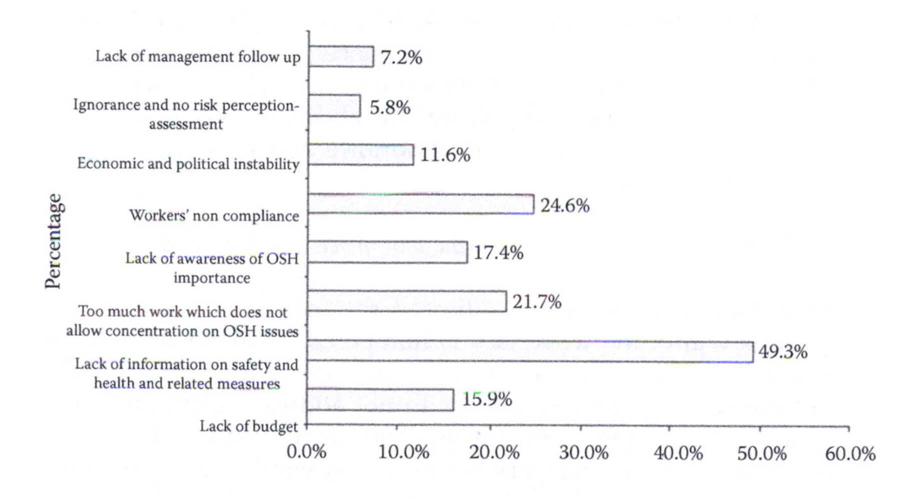

Figure 8.9 Factors impeding enterprises from providing OSH.

- Over 50% of respondents indicated that OSH improves the image of the company.
- Around 55% of the respondents indicated that OSH denotes ethics-care for workers.
- 29.0% and 8.8% felt that introducing OSH at the workplace is a sign of respect for international safety standards and national OSH regulations, respectively.

Factors impeding enterprises from providing OSH (see Figure 8.9):

- Close to 50% of respondents indicated that lack of information on safety, health and related measures impede enterprises from

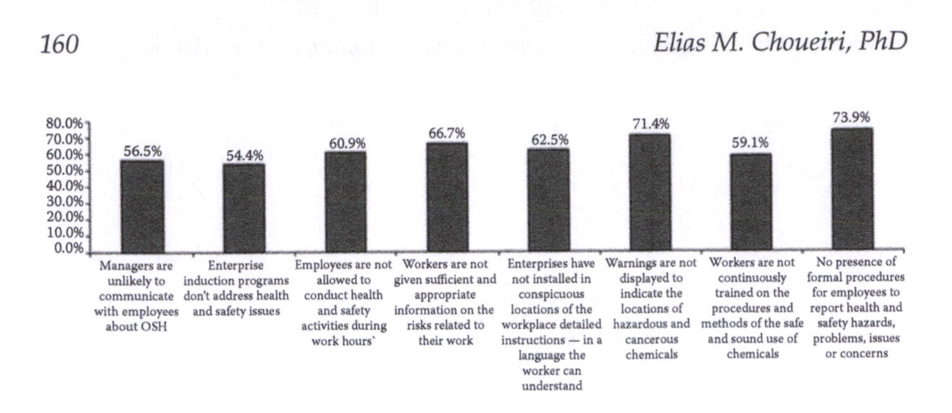

Figure 8.10 Practices relating to safety and health.

providing OSH, followed by workers' non-compliance (24.6%), too much work which does not allow concentration on OSH issues (21.7%), lack of awareness of OSH importance (17.4%), lack of budget (15.9%), economic and political instability (11.6%), lack of management follow-up (7.2%), and ignorance and no-risk perception assessment (5.8%);

Communication of OSH policies within the workplace (see Figure 8.10):

- Over 54% of respondents indicated that communication of OSH policies within the workplace is hampered by managers who are unlikely to communicate with employees about OSH (56.5%); enterprises induction programs do not address health and safety issues (54.4%); employees are not allowed to conduct health and safety activities during work hours (60.9%); employees are not given sufficient and appropriate information on the risks related to their work (66.7%); enterprises do not install in conspicuous locations of the workplace detailed instructions in a language that the worker can understand (62.5%); warnings are not displayed to indicate the locations of hazardous and cancerous chemicals (71.4%); workers are not continuously trained on the procedures and methods of the safe and sound use of chemicals (59.1%); no presence of formal procedures to report health and safety hazards, problems, issues or concern (73.9%).

Relations and communication of enterprises with the Ministry of Labour (see Figure 8.11):

- 100% of enterprises indicated that they did not notify the Ministry of Labour of occupational accident within 24 hours after their occurrence.

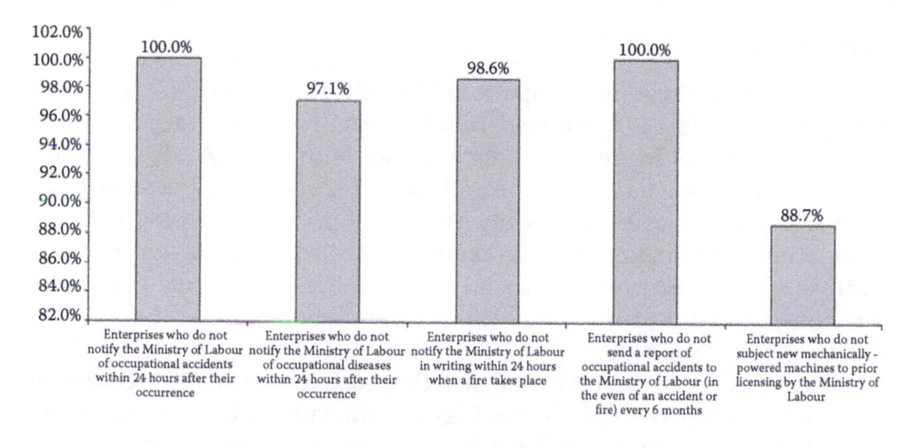

Figure 8.11 Communication with the Ministry of Labour.

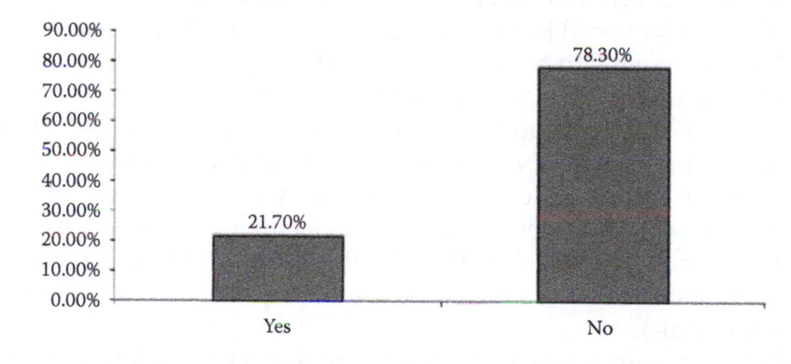

Figure 8.12 Accident documentation.

- 97.1% of enterprises indicated that they did not notify the Ministry of Labour of occupational diseases within 24 hours after their occurrence.
- 98.6% of enterprises indicated that they did not notify the Ministry of Labour in writing within 24 hours after a fire had taken place.
- 100% of enterprises did not send a report on occupational accidents to the Ministry of Labour every 6 months.
- 88.7% of the enterprises indicated that they did not subject new mechanically powered machines to prior licensing by the Ministry of Labour.

Accident documentation (see Figure 8.12):

- Only 78.3% of enterprises indicated that they document work-related accidents.

Conclusions

The primary obstacle to occupational health in most MENA countries remains the lack of a political mechanism that translates information into action. In reality, policy makers in MENA do not lack information. A casual walk through any type of workplace in most MENA countries would easily uncover a range of unsafe practices and occupational hazards. Policy makers are still driven by the need to address other 'more pressing' social and health issues that are politically less complicated and more saleable to the general public (Nuwayhid, 2004).

The solution to occupational health problems in MENA countries therefore requires not only technological innovation but also significant institutional and legal developments. Occupational health researchers should understand the 'political economy' of the labour market at global, regional, and nation–state levels. They must recognize the leading role of forces fighting for social justice, particularly the role of organised labour, which is instrumental to advancing national occupational health agendas and ratifying international labour laws, notwithstanding the repression they face and their questionable representation of the interest of their constituency in many MENA countries. Occupational health researchers in MENA also must be alert to the potentially negative effect of global trade on the health and safety of poor and marginalised workers. Research should contribute to the international call to hold multinational corporations accountable to international ethical occupational health practices (Nuwayhid, 2004).

Employers everywhere in the MENA region have a safety obligation. They must take the necessary and appropriate measures to protect workers' health and prevent occupational risk. Employers must likewise comply with the law in respect of recompense for harm or loss arising from accidents at work and occupational diseases.

References

A Note on Disability Issues in the Middle East and North Africa. (2005). Human Development Department Middle East and North Africa Region, The World Bank.

Armour, M. A. (2003). *Hazardous Laboratory Chemicals and Disposal Guide.* CRC Press.

Azzi, M. (2009). Challenges in incorporating safety and health into national plans: Between policy and practice in Lebanon. *International ILO Conference, Düsseldorf, Germany,* 3–6 November 2009.

Baer, M. and Frese, M. (2003). Innovation is not enough: Climates for Initiative and Psychological Safety, Process Innovation and Firm Performance. *Journal of Organizational Behaviour*, 24, 45–68.

Dimitracopoulos, A. (2008). Construction contracting in the Middle East: Regional departures from international practices. *Construction Law International*, 3(1), 6–8.

Gillen, M., Baltz, D., Gassel, M. Kirsch, L. and Vaccaro, D. (2002). Perceived safety climate job demand and worker support among union and non union injured construction workers. *Journal of Safety Research*, 33(1), 33–51.

Gwandure, C. and Thatcher, A. (2006). The impact of voluntary withdrawal from anti-epileptic medication on job performance and workplace safety. *Ergonomics SA*, 18(2), 30–43.

Hatch-Maillette, M. A. and Scalora, M. J. (2002). Gender, sexual harassment, workplace violence and risk assessment: Convergence around psychiatric staffs' perceptions of personal safety aggression and violent behaviour. *Science Direct*, 7, 271–291.

Henshaw, J. L., Gaffney, S. H., Madl, A. K. and Paustenback, D. (2007). The employer's responsibility to maintain a safe and healthful work environment: A historical review of societal expectations and industrial practices. *Employee Response Right Journal*, 19, 173–192.

Imperative Need for Occupational Health and Safety in Middle East. (2012). *Fleming Gulf Conferences, Middle East OHS Strategy Summit*, 5–7 March 2012, Abu Dhabi.

Iunes, R. (n.d.). Occupational Safety and Health in Latin America and the Caribbean: Overview, Issues and Policy Recommendations. Operational Department 3, Inter-American Development Bank.

Mehta, P. K. and Burrows, R. W. (2001). Building durable structures in the 21st century. *Concrete International*, 23(3), 57–63.

Miller, N. (2001). Stalking Laws and Implementation Practices: A National Review of Policymakers and Practitioners. *Institute of Law and Justice Research, Social Science Research Network.*

Mzid, N. (2004). *Comparative Study of Labour Law in the Arab Partner Countries.* Brussels, Belgium: European Trade Union Institute.

Nuwayhid, I. (2004). Occupational health research in developing countries: A partner for social justice. *Journal of Public Health*, 94(11), 1916–1921

Occupational Safety and Health (2011). International Labour Organization.

Overview of the Occupational Safety and Health Situation In the Arab Region. (2007). *Inter-Regional Tripartite Meeting on Occupational Safety and Health, International Labour Organization, 18–20 November 2007, Damascus, Syria.*

Profumo, A., Spini, G., Cucca, L. and Pesavento, M. (2003). Determination of inorganic nickel compounds in the particulate matter of emissions and workplace air by selective sequential dissolutions. *Talanta*, 61(4), 465–472.

Russeau, S. (2011). The Plight of Migrant Workers in the Middle East and North Africa. *Global Research*. Retrieved from http://www.globalresearch.ca/the-plight-of-migrant-workers-in-the-middle-east-and-north-africa/24879

Saudi Arabic to Lead MENA Construction Boom (2012). Retrieved from http://www.constructionweekonline.com/

Valent, F., McGwin, G., Bovenzi, M. and Barbone, F. (2002). Fatal work-related inhalation of harmful substances in the United States. *Chest Journal*, 121(3), 969–975.

Zwetsloot, G. (2003). From management system to corporate social responsibility. *Journal of Business Ethics*, 44.

Appendix

Contractor safety, health and environmental regulations

Lebanon

Table of clauses

1. Introduction

1.1 The prevention of injury and/or illness to site personnel and the public, damage to the Works and to public and private property, protection of the environment, and compliance with applicable laws, are primary objectives of the Council for Development and Reconstruction (CDR) (the Employer). Because of the importance CDR places on meeting these objectives, selected minimum requirements are outlined in these Safety, Health and Environmental Regulations with which Contractors shall comply whilst working on CDR contracts. Given that these Regulations cannot cover every eventuality, the Contractor shall be expected to exercise good judgment in all such matters, even though not mentioned in these Regulations, and shall take any and all additional measures, as required or necessary, to meet his responsibility for safety, health and environmental matters during the period of the Contract.

CDR and its representatives shall not be held liable for any actions taken by the Contractor that are attributed to following the minimum requirements stated hereinafter.

1.2 The Contractor shall, throughout the execution and completion of the Works and the remedying of any defects therein:

(a) have full regard for the safety of all persons on the Site and keep the Site and the Works in an orderly state appropriate to the avoidance of danger to any person;

(b) know and understand all laws governing his activities along with any site requirements and work site hazards. Such information shall be communicated by the Contractor to his personnel and subcontractors;

(c) take all necessary measures to protect his personnel, the Employer's personnel, other persons, the general public and the environment;

(d) avoid damage or nuisance to persons or to property of the public or others resulting from pollution, noise or other causes arising as a consequence of carrying out the Works;

(e) ascertain and comply with any regulations concerning noise, pollution and other nuisance in addition to the obligations imposed by the Conditions of Contract and by law;

(f) take necessary precautions to prevent nuisance from smoke, dust, rubbish, water, polluted effluent and other causes.

2. *Compliance with regulations*

2.1 The Contractor shall comply with the requirements of these Safety, Health and Environmental Regulations and all other applicable regulations or requirements under Lebanese laws, laid down by relevant authorities or issued by the Employer or the Engineer concerning safety, health and the environment, in force or introduced or issued from time to time during the period of the Contract.

In so far as these Regulations are applicable, they shall apply to sites and personnel outside the Site associated with the performance of the Contract.

2.2 The Regulations equally apply to subcontractors and all other parties engaged by the Contractor and their personnel. The Contractor shall ensure all such parties are fully aware of and comply with the Regulations.

2.3 The Contractor shall comply with all notifications and written or verbal instruction regarding safety issued pursuant to these Regulations by the Employer, Engineer or relevant authorities within the time specified in the notification or instruction.

Whenever the Contractor is required to obtain the approval, agreement, permission, etc. of the Engineer, such approval, agreement, permission, etc. shall not relieve the Contractor of his responsibilities and obligations under these Regulations or the Contract.

2.4 The Contractor shall adopt a positive approach, awareness and responsibility towards safety, health and the environment, and take appropriate action, by:

(a) ensuring the Regulations are enforced and followed by the Contractor's personnel. Any failure by the Contractor's personnel to follow the Regulations, shall be regarded as a failure by the Contractor.

(b) paying attention to possible injury to unauthorised persons entering the site, particularly children

2.5 Whenever in these Regulations the Contractor is required to provide test certificates for equipment and personnel or to comply the relevant authorities' requirements and no independent test facilities are available or no relevant authorities exist in Lebanon, the Contractor shall provide:

(a) in lieu of independent test certificates:

for equipment—details of the tests and the date of the tests that have been carried out by the Contractor and a written statement that the Contractor has satisfied himself that the item of equipment is fit and safe for use;

for personnel—details of the training and experience and a written statement that the Contractor has satisfied himself that the person has the required level of competency.

(b) in lieu of relevant authorities' requirements—details of the Contractor's own rules, regulations, requirements and procedures regarding safety, health and the environment.

If the Engineer is dissatisfied with the details provided by the Contractor, the Contractor shall provide further details or carry out further tests or provide further written statements as may be reasonably required by the Engineer.

When the Engineer has satisfied himself regarding the Contractor's own rules, regulations, requirements and procedures provided in accordance with (b) above, such rules, etc. shall be deemed to form part of these Regulations and to which Clause 3 shall equally apply.

3. *Failure to comply with regulation*

3.1 General

3.1.1 Should the Contractor fail to comply with any of the Regulations or requirements:

(a) the Engineer may suspend the Works or part of the Works until the Contractor has taken necessary steps, to the satisfaction of the Engineer, to comply with the regulations or requirements.

(b) the Employer may, following written notice to the Contractor, carry out themselves or arrange for another contractor to carry out such measures as they consider appropriate on behalf of the Contractor. Any such actions by the Employer shall not affect or diminish the Contractor's obligations or responsibilities under the Contract.

(c) the Engineer may, following written notice to the Contractor, deduct from payments to the Contractor the amounts stipulated in Sub-Clause 3.2. Such notice shall specify:
   (i) the nature of the failure or failures;
   (ii) the period after the date of the notice within which the Contractor shall remedy each failure; and
   (iii) the amount to be deducted.

Such suspension of payment will remain in force until such time as the Contractor has rectified the breach or breaches to the satisfaction of the Engineer. No interest shall be paid on the suspended payments.

3.1.2 Failure to comply with the Regulations or requirements shall be considered a breach of contract by the Contractor and may result in termination of the Contract by the Employer.

3.1.3 In the event of the Employer or Engineer taking action based on Sub-Clause 3.1.1(a) or (b) or 3.1.2, the Contractor shall not be entitled to any additional costs or extension to the Contract Completion Date.

3.1.4 All costs incurred by the Employer pursuant to Sub-Clause 3.1.1(b) and the deductions from payments imposed on the Contractor by the Engineer under Sub-Clause 3.1.1(c) shall be deducted from amounts otherwise due to the Contractor.

3.2 Deductions from Payments

3.2.1 Failures by the Contractor to comply with the Regulations or requirements are classified as follows:
D1—breaches of Sub-Clause 5.6 (personal protective equipment);
D2—breaches of Clause 7 (work in Public Areas);
D3—breaches other than D1 and D2.

3.2.2 The basic deduction from payment for each classification in Sub-Clause 3.2.1, is as follows:
for D1—USD 100/person/day;
for D2—USD 500/location/day;
for D3—USD 100/occurrence/day.
Limit of cumulative total deductions shall not exceed 1% of Contract price.

3.2.3 Deductions from payments will be applied as follows:
(a) for the first breach of each regulation or requirement—the basic deduction. If the same or similar breaches occur in different situations or locations at the same time, the Engineer may apply deductions for each situation or location; this will not apply to breaches related to personal protective equipment.

(b) for a second or subsequent breach of the same Regulation or requirement or failure to rectify a previous failure within the time specified by the Engineer—twice the basic deduction.

4. General requirements

4.1 Preamble

4.1.1 All references to safety shall be deemed to include health and the environment.

4.2 Safety Officer

4.2.1 The Contractor shall appoint a competent Safety Officer who shall be responsible for safety, health and the environment. The Safety Officer shall be given sufficient time by the Contractor to carry out his duties; minimum requirements shall be as follows:

Workforce on Site of over 250—full time Safety Officer;

Workforce on Site of 100–250—50% of Safety Officer's time;

Workforce on Site below 100—as required for the Works but a minimum of 5 hours per week of Safety Officer's time where more than 20 workers.

4.2.2 The Contractor shall provide the Safety Officer with appropriate identification, including a white hard hat with red cross symbol and an identification badge. The appointment of the Safety Officer shall be in writing and copied to the Engineer. The appointment shall include specific instructions to enforce these Regulations and delegated authority to take any action, measure or to issue instructions regarding their enforcement. All persons on Site shall be made aware of the name and authority of the Safety Officer and instructed to comply with any instruction or direction on safety matters, verbal or in writing, issued by the Safety Officer.

4.2.3 The Safety Officer shall be provided with a mobile phone or other similar means of communication. The Safety Officer shall be accessible and available at all times including outside normal working hours.

4.3 Safety Training

4.3.1 The Contractor shall provide safety induction training for all site personnel upon starting on site.

4.3.2 The Contractor shall provide safety refresher/reinforcement training at regular intervals for his staff.

4.4 Safety Meetings

4.4.1 The Contractor shall hold regular safety meetings to provide safety instructions and receive feedback from site personnel on safety, health and environmental matters. A weekly Safety Meeting shall be chaired by the Safety Officer and minutes shall be taken of the meeting. The meeting/minutes shall cover all relevant issues including actions to be taken. A copy of the minutes shall be given to the Engineer. The Safety Officer should attend the Contractor's weekly site meetings and 'Safety' should be an item on the agenda.

4.5 Safety Inspections

4.5.1 The Safety Officer shall make regular safety inspections of the work site. The Safety Officer shall prepare a report of each inspection. This report shall include details of all breaches of these Regulations and any other matters or situations relating to safety found during the inspection, instructions issued by the Safety Officer and actions taken by the Contractor. A copy of the Safety Officer's inspection reports shall be given to the Engineer.

4.6 Control of Substances Hazardous to Health

4.6.1 Hazardous materials shall be stored in approved safety containers and handled in a manner specified by the manufactures and/or prescribed by relevant Authorities (see Sub-Clause 2.5).

4.6.2 Only properly trained and equipped personnel shall handle hazardous materials.

4.7 Potential Hazards

4.7.1 The Contractor shall inform employees of potential hazards, take appropriate steps to reduce hazards and be prepared for emergency situations.

4.7.2 The Contractor shall make an assessment of every operation involving hazardous substances. The assessment shall be recorded on a Hazardous and Flammable Substances Assessment Method Statement which shall be submitted to the Engineer prior to the delivery and use of the substance on Site.

4.8 Accident Reporting

4.8.1 The Contractor shall report all accidents and dangerous occurrences to the Engineer. The Contractor shall prepare a report on each accident or dangerous occurrence and a copy of the report, together with witness statements and any other relevant information, shall be submitted to the Engineer. A reportable accident or dangerous occurrence shall include any accident to any person on Site requiring medical attention or resulting in the loss of working hours or any incident that resulted, or could have resulted, in injury, damage or a danger to the Works, persons, property or the environment.

4.8.2 In the event of an accident or dangerous occurrence, the Contractor shall be responsible for completing all statutory notifications and reports. Copies of all statutory notifications and reports shall be passed to the Engineer.

4.8.3 All accidents and dangerous occurrences shall be recorded in a Site Accident Book. The Site Accident Book shall be available at all times for inspection by the Engineer.

4.8.4 The Contractor shall immediately rectify any situation or condition that could result in injury, damage or a danger to the Works, person, property or the environment. If the situation or condition cannot be corrected immediately, the Contractor shall provide temporary barriers and appropriate warning signs and devices and/or take other appropriate action necessary for the protection of persons, property and the environment.

4.9 Notices, Signs, Etc.

4.9.1 All safety, health, environmental and other notices and signs shall be clearly displayed and written in both Arabic and either English or French. All requirements, instructions, procedures, etc. issued by the Contractor concerning these Regulations shall be printed in both Arabic and English and displayed and readily available to Contractor's personnel.

4.10 First Aid and Medical Attention

4.10.1 The Contractor shall have comprehensive First Aid Kit(s) on Site at all times. First Aid Kits shall be conveniently located and clearly identifiable.

4.10.2 The Contractor shall have one employee on site trained in first aid for every 25 employees. Such persons shall be provided with appropriate identification, including a red hard hat with a white 'red cross' symbol and an identification badge.

4.10.3 The Contractor shall make contingency arrangements for calling a Doctor and transporting injured persons to hospital. The telephone numbers of the emergency services and the name, address and telephone number of the Doctor and nearest hospital shall be prominently displayed in the Contractor's site office.

4.11 Employee Qualifications and Conduct

4.11.1 The Contractor shall employ only persons who are fit, qualified and skilled in the work to be preformed. All persons shall be above the minimum working age.

4.11.2 Contractor's personnel shall use the toilet facilities provided by the Contractor. 4.11.3 The Contractor shall ensure:

(a) that no firearms, weapons, controlled or illegal substances or alcoholic beverages are brought onto the Site and that no personnel under the influence of alcohol or drugs are permitted on Site.

(b) that all personnel obey warning signs, product or process labels and posted instructions.

(c) that drivers or operators of vehicles, machinery, plant and equipment follow the rules for safe operations. Drivers shall wear seat belts and obey all signs and posted speed limits.

4.12 Security

4.12.1 The Contractor shall take all measures necessary, including watching and lighting at night, to prevent unauthorised entry to the Site and to safeguard the Site, the Works, materials, Plant, Contractor's Equipment and Temporary Works against damage from trespass and theft.

5. Safety requirements

5.1 Personal Protective Equipment

5.1.1 The Contractor shall provide personal protective equipment, including hard hats, safety glasses, respirators, gloves, safety shoes, and such other equipment as required, and shall take all measures or actions for the protection and safety of Contractor's personnel.

5.1.2 Non-metallic hard hats shall be worn at all times by all personnel at the worksite with the exception of those areas where the Engineer has indicated it is not necessary to do so.

5.1.3 Safety glasses shall meet international standards and be available for use and worn in specified worksite areas. As a minimum, safety glasses shall be worn for the following types of work: hammering, chipping, welding, grinding, use of electrically powered or pneumatic equipment, insulation handling, spray painting, working with solvents, and other jobs where the potential of an eye injury exists. Face shields and/or monogoggles shall be worn where possible exposure to hazardous chemicals, cryogenic fluids, acids, caustics, or dust exists and where safety glasses may not provide adequate protection.

5.1.4 When handling acids, caustics, and chemicals with corrosive or toxic properties, suitable protection, such as acid suits or chemical resistant aprons and gloves, shall be worn to prevent accidental contact with the substance.

5.1.5 Personnel shall not be permitted to work whilst wearing personal clothing or footwear likely to be hazardous to themselves or others.

5.1.6 The wearing of safety shoes with steel reinforced toes is recommended for all Contractor's personnel on site. In all cases, Contractor's personnel shall wear substantial work shoes that are commensurate with the hazards of the work and the worksite area.

5.1.7 Hearing protection, including muffs, plugs or a combination thereof, shall be provided for all personnel operating in areas where the noise level exceeds 90 decibels. Such protection shall also be provided for operators working with equipment exceeding such a level. This may include equipment such as excavators, shovels, jackhammers, saws, drills, grinders, and the like are being used.

5.1.8 The Contractor shall encourage employees to wear substantial work gloves whenever practical and safe to do so.

5.2 Fire Protection and Prevention

5.2.1 The Contractor shall take all necessary measures to prevent personal injury or death or damage to the Works or other property, including but not limited to:

(a) provision of fire fighting facilities in all vulnerable areas and as instructed by the Engineer;

(b) marking escape routes and illuminating them if necessary;

(c) instructing workmen in fire precautions and use of fire fighting equipment;

(d) displaying notices on fire safety and procedures in the event of a fire on Site.

5.2.2 The Contractor shall comply with fire protection instructions given by the Authorities having jurisdiction in regard to fire protection regulations.

5.2.3 The Contractor shall, upon moving on site, provide to the Engineer and the Authorities a fire prevention and evacuation plan. This shall include drawing(s) showing the fire assembly points. The fire prevention and evacuation plan and drawing(s) shall be updated from time to time as the Works progress. The Contractor shall ensure all personnel are fully informed on escape routes and assembly points and any changes thereto.

5.2.4 Fuel storage will not be permitted in construction work areas. Contractors may establish fuel storage tanks in special areas set aside for the purpose and approved by the Engineer. Storage tanks shall be adequately bounded to control spillage. Fire extinguishers shall be provided and installed in a suitable nearby location.

5.2.5 Highly combustible or volatile materials shall be stored separately from other materials and as prescribed by relevant authorities and under no circumstances within buildings or structures forming part of the permanent Works. All such materials shall be protected and not exposed to open flame or other situations which could result in a fire risk.

5.2.6 No combustible site accommodation shall be located inside or within 10 meters of a building or structure forming part of the permanent Works, where units have to be used in these circumstances, they shall be constructed of non-combustible materials and have a half-hour fire rating inside to outside and outside to inside. Non-combustible furniture shall be used where practical.

5.2.7 All temporary accommodation and stores shall be provided with smoke detectors and fire alarms.

5.2.8 Smoking shall be banned in high-risk areas.

5.2.9 Expanded polystyrene with or without flame retarding additive, polythene, cardboard and hardboard shall not be used as protection materials.

5.2.10 Plywood and chipboard shall only be used as protection on floors. Vertical protection shall be non-combustible. Debris netting and weather protection sheeting shall be fire retardant.

5.2.11 When using cutting or welding torches or other equipment with an open flame, the Contractor shall provide a fire extinguisher close by at all times. All flammable material shall be cleared from areas of hot works, or work locations prior to welding or oxy/gas burning operations. All hot works shall cease half an hour before the end of a work shift to allow for thorough checking for fires or smouldering materials. Where appropriate, areas of hot works are to be doused in water before the shift ends.

5.2.12 An adequate number of fire extinguishers of types suited to the fire risk and the materials exposed shall be provided. These shall be placed in accessible, well-marked locations throughout the job site. Contractor's personnel shall be trained in their use. Extinguishers shall be checked monthly for service condition and replaced or recharged, as appropriate after use.

5.2.13 Only approved containers shall be used for the storage, transport and dispensing of flammable substances. Portable containers used for transporting or transferring gasoline or other flammable liquids shall be approved safety cans.

5.2.14 Fuel burning engines shall be shut off whilst being refuelled.

5.2.15 Adequate ventilation to prevent an accumulation of flammable vapours shall be provided where solvents or volatile cleaning agents are used.

5.2.16 Flammables shall not be stored under overhead pipelines, cable trays, electrical wires, or stairways used for emergency egress.

5.2.17 Paints shall be stored and mixed in a room assigned for the purpose. This room shall be kept under lock and key.

5.2.18 Oily waste, rags and any other such combustible materials shall be stored in proper metal containers with self-closing lids and removed every night to a safe area or off site. Every precaution shall be taken to prevent spontaneous combustion.

5.3 Electrical Safety

5.3.1 All temporary electrical installations, tools and equipment shall comply with current regulations dealing with on-site electrical installations.

5.3.2 The Contractor shall establish a permit-to-work system for work on or in proximity to energised circuits of any voltage. Contractor's personnel shall not commence work on such circuits unless a permit to work has been issued and adequate safety measures have been taken and the work operation has been reviewed and approved by the Engineer.

5.3.3 Only authorised personnel shall be allowed to work or repair electrical installations and equipment.

5.3.4 Portable tools and equipment shall be 220 volt, unless otherwise agreed by the Engineer.

5.3.5 When portable or semi-mobile equipment operates at voltages in excess of 110 volts, the supply shall be protected by a Residual Current Device (RCD) regardless of any such device fitted to the equipment. The RCD must have a tripping characteristic of 30 milliamps at 30 milliseconds maximum.

5.3.6 All static electrically powered equipment, including motors, transformers, generators, welders, and other machinery, shall be properly earthed, insulated, and/or protected by a ground fault interruption device. In addition, the skin of metal buildings and trailers with electric service shall be earthed. Metal steps, when used, shall be securely fixed to the trailer.

5.3.7 Lamp holders on festoon lighting shall be moulded to flexible cable and be of the screw in type. Clip on guards shall be fitted to each lamp unit.

5.3.8 All tungsten-halogen lamps shall be fitted with a glass guard to the element. These lamps must be permanently fixed at high level.

5.3.9 Electrical equipment shall be periodically inspected and repaired as necessary by competent persons.

5.3.10 Any work on electrical equipment and systems shall be made safe through locking, tagging, and/or isolation of the equipment before work commences. Prior to the start of the work, the equipment or systems shall be tested to insure that they have been properly de-energised and isolated.

5.3.11 Electrical repair work on energised systems shall be avoided whenever possible.
5.3.12 Electrical troubleshooting shall be conducted only after getting written approval of the Engineer.
5.3.13 Unauthorised personnel shall not enter enclosures or areas containing high voltage equipment such as switchgear, transformers, or substations.

5.4 Oxygen/Acetylene/Fuel Gases/Cartridge Tools
5.4.1 Compressed oxygen shall never be used in the place of compressed air.
5.4.2 Flash-back (Spark) arrestors shall be fitted to all gas equipment.
5.4.3 Liquid Petroleum Gas (LPG) cylinders shall not be stored or left in areas below ground level overnight. Cylinders must be stored upright.
5.4.4 The quantity of oxygen, acetylene and LPG cylinders at the point of work shall be restricted to a maximum of one day's supply. Cylinders shall be kept in upright vertical rack containers or be safely secured to a vertical support.
5.4.5 Cartridge tools shall be of the low velocity type. Operators must have received adequate training in the safe use and operation of the tool to be used.

5.5 Scaffolding/Temporary Works
5.5.1 No aluminium tube shall be used, except for proprietary mobile towers, unless otherwise agreed with the Engineer.
5.5.2 Drawings and calculations shall be submitted to the Engineer, prior to commencement of work on site, for all Temporary Works, including excavations, falsework, tower cranes, hoists, services and scaffolding. Design shall conform to international standards.
5.5.3 The Engineer will not approve Temporary Work designs but the Contractor shall take account of any comments on such designs made by the Engineer.
5.5.4 The Contractor shall inspect and approve all Temporary Works after erection and before access, loading or use is allowed. Completed and approved Temporary Works shall be tagged with a scaff-tag or similar safety system and the Safe Structure insert displayed. For scaffolding, one tag shall be displayed every 32 m^2 of face area. A central record system shall be kept on all Temporary Work. Temporary Works shall be inspected weekly and similarly recorded.
5.5.5 All mobile scaffold towers shall be erected in accordance with the manufacture's instructions and a copy of these shall be submitted to the Engineer prior to any use on site.

Additionally, all towers shall be erected complete with access ladder, safety rails and kick boards whatever the height.

5.5.6 The Contractor shall repair or replace, immediately, any scaffold including accessories, damaged or weakened from any cause.

5.5.7 The Contractor shall ensure that any slippery conditions on scaffolds are eliminated as soon as possible after they occur.

5.5.8 All scaffolds used for storing materials, for brick or block laying, for access to formwork or for any other purpose where materials may accidentally fall, shall be provided with wire mesh guards or guards of a substantial material, in addition to kick boards.

5.6 Use of Ladders

5.6.1 Manufactured ladders shall meet the applicable safety codes for wood or metal ladders. Metal ladders shall not be used where there is any likelihood of contract with electric cables and equipment. All metal ladders shall be clearly marked: 'Caution—Do not use around electrical equipment'.

5.6.2 Job made ladders shall not be permitted.

5.6.3 Extension or straight ladders shall be equipped with non-skid safety feet, and shall be no more than 12 m in height. The maximum height of a stepladder shall be 2 m. Ladders shall not be used as platforms or scaffold planks.

5.6.4 Ladders rungs and steps shall be kept clean and free of grease and oil.

5.6.5 Extension and straight ladders shall be tied off at the top and/or bottom when in use. Only one person shall be allowed on a ladder at a time.

5.6.6 Defective ladders shall be taken out of service and not used. Ladders shall not be painted and shall be inspected for defects prior to use.

5.7 Elevated Work

5.7.1 The Contractor shall provide all personnel, whilst working at an elevated position, with adequate protection from falls. Details of such protection shall be submitted to and approved by the Engineer.

5.7.2 The Contractor shall carry out daily inspections of all elevated work platforms. Defects shall be corrected prior to use.

5.7.3 Roofing and Sheet Material Laying

(a) A Method Statement detailing the procedures to be adopted shall be submitted to and agreed with the Engineer prior to commencement of work on site.

(b) Mobile elevating work platforms or the equivalent shall be used to install roofing and sheet materials wherever practicable and a suitable base is available.

5.7.4 Erection of Structures
(a) A Method Statement detailing the procedures to be adopted shall be submitted and agreed with the Engineer prior to commencement of work on site.
(b) Safety harnesses and lines shall be provided by the Contractor for use by the erection personnel and worn at all times.
(c) Mobile elevating work platforms or the equivalent shall be used to erect structures wherever practicable and a suitable base is available.

5.7.5 Mobile Elevating Work Platforms
Operators shall be trained in the safe use of such platforms and hold a current Certificate of Competence (see Sub-Clause 2.5).

5.7.6 Hoists
(a) A copy of the current Test Certificate (see Sub-Clause 2.5) shall be submitted to the Engineer before any hoist (personnel or material) is brought into operation on the site. Where the range of travel is increased or reduced a copy of the revised Test Certificate shall be submitted.
(b) Each landing gate shall be fitted with a mechanical or electrical interlock to prevent movement of the hoist when any such gate is in the open position.
(c) Safety harnesses must be worn and used by personnel erecting, altering and dismantling hoists.

5.7.7 Suspended Cradles
(a) Suspended cradles shall be installed, moved and dismantled by a specialist contractor.
(b) Suspended cradles shall comply with local regulations.
(c) All powered suspended cradles shall incorporate independent safety lines to overspeed braking devices and independent suspension lines for personal safety harness attachment.

5.8 Use of Temporary Equipment

5.8.1 The safe design capacity of any piece of equipment shall not be exceeded, nor shall the equipment be modified in any manner that alters the original factor of safety or capacity.

5.8.2 Mobile equipment shall be fitted with suitable alarm and motion sensing devices, including backup alarm, when required.

5.8.3 The Contractor shall ensure that the installation and use of equipment are in accordance with the safety rules and recommendations laid down by the manufacturer, taking into account the other installations already in place or to be installed in the future.

5.8.4 The Contractor shall inspect Equipment prior to its use on the Works and periodically thereafter to ensure that it is in safe working order. Special attention shall be given to such items as cables, hoses, guards, booms, blocks, hooks and safety devices. Equipment found to be defective shall not be used and immediately removed from service, and a warning tag attached.

5.8.5 Natural and synthetic fibre rope made of material such as manila, nylon, polyester, or polypropylene shall not be used as slings if approved by the Engineer.

5.8.6 Only trained, qualified and authorised personnel shall operate equipment. All drivers and operators shall hold a current Certificate of Training Achievement for the equipment being used (see Sub-Clause 2.5).

5.8.7 A safety observer shall be assigned to watch movements of heavy mobile equipment where hazards may exist to other personnel from the movement of such equipment, or where equipment could hit overhead lines or structures. The observer shall also ensure that people are kept clear of mobile equipment and suspended loads.

5.8.8 When mobile or heavy equipment is travelling onto a public thoroughfare or roadway, a flagman shall insure that traffic has been stopped prior to such equipment proceeding. Whilst the mobile or heavy equipment is travelling on a public roadway, a trailing escort vehicle with a sign warning of a slow-moving vehicle that is dangerous to pass shall be provided.

5.8.9 Cranes:

(a) The Contractor shall give a minimum of 48 hours notice to the Engineer prior to bringing a mobile crane on site.

(b) No cranes shall be erected on the site without the prior approval of the Engineer. The Engineer may direct the Contractor as to locations where cranes may not be located. The Contractor shall take such directions into account when submitting his proposals for crane location points, base footings, pick up points and swing radius. Compliance with any such direction shall not entitle the Contractor to any extension of the Period of Completion or to any increase in the Contract Price.

(c) Safety harnesses shall be worn and used at all times by personnel engaged on the erection, alterations and dismantling of tower cranes.

(d) The Contractor shall provide a copy of the current Test Certificate (see Sub-Clause 2.5) to the Engineer before any crane (tower or mobile) is brought into operation on the Site.

(e) All lifting tackle must hold a current Test Certificate (see Sub-Clause 2.5). All lifting tackle must be thoroughly examined every 6 months and an inspection report raised.

(f) All fibrous/web slings shall be destroyed and replaced 6 months after first use.

(g) All crane drivers/operators shall hold a Certificate of Training Achievement for the class of crane operated (see Sub-Clause 2.5).

(h) All banksmen/slingers shall hold a Training Certificate from a recognised training agency (see Sub-Clause 2.5).

(i) Only certified slingers/banksmen shall sling loads or guide crane/load movement.

(j) The maximum weekly working hours of a crane driver or banksman shall be restricted to 60 hours.

(k) Under no circumstances, shall a crane or load come within 4 m of any energised overhead power line or other critical structure.

5.9 Locking-Out, Isolating, and Tagging of Equipment

5.9.1 Equipment that could present a hazard to personnel if accidentally activated during the performance of installation, repair, alteration, cleaning, or inspection work shall be made inoperable and free of stored energy and/or material prior to the start of work. Such equipment shall include circuit breakers, compressors, conveyors, elevators, machine tools, pipelines, pumps, valves, and similar equipment.

5.9.2 Where equipment is subject to unexpected external physical movement such as rotating, turning, dropping, falling, rolling, sliding, etc., mechanical and/or structural constraints shall be applied to prevent such movement.

5.9.3 Equipment which has been locked-out, immobilised, or taken out of service for repair or because of a potentially hazardous condition shall be appropriately tagged indicating the reason it has been isolated and/or taken out of service.

5.9.4 Where safety locks are used for locking out or isolating equipment, the lock shall be specially identified and easily recognised as a safety lock.

5.10 Installation of Temporary or Permanent Equipment

5.10.1 During installation and testing the Contractor's specialist engineer shall be in attendance.

5.10.2 All control mechanism panel and wiring diagrams shall be available and printed in both Arabic and either English or French.

5.11 Laser Survey Instruments

5.11.1 Details of the types and use of laser instruments shall be submitted and agreed with the Engineer.

5.12 Working in Confined Spaces

5.12.1 Confined spaces, including tanks, vessels, containers, pits, bins, vaults, tunnels, shafts, trenches, ventilation ducts, or other enclosures where known or potential hazards may exist, shall not be entered without prior inspection by and authorization from the Site Safety Officer and the issuance of a Hazardous Work Permit.

5.12.2 Prior to entering the confined space, the area shall be completely isolated to prevent the entry of any hazardous substances or materials which could cause an oxygen deficient atmosphere. All equipment that could become energised or mobilised shall be physically restrained and tagged. All lines going into the confined space shall be isolated and/or blanked.

5.12.3 Personnel working in a confined space where emergency escape or rescue could be difficult, shall wear a safety harness attached to a lifeline.

5.12.4 A qualified attendant(s), trained and knowledgeable in job-related emergency procedures, shall be present at all times whilst persons are working within the confined space. The attendant shall be capable of effecting a rescue, have necessary rescue equipment immediately available, and be equipped with at least the same protective equipment as the person making entry.

5.12.5 All equipment to be used in a confined space shall be inspected to determine its acceptability for use. Where a hazard from electricity may exist, equipment utilised shall be of low voltage type.

5.12.6 The atmosphere within the confined space shall be tested to determine it is safe to enter. Acceptable limits are:

—oxygen: 19.5% lower, 22% higher;

—flammable gas: not to exceed 10% of lower explosion limit;

—toxic contaminants: not to exceed the permissible exposure limit.

Subsequent testing shall be done after each interruption and before re-entering the confined space, as well as at intervals not exceeding 4 hours. Continuous monitoring is preferable and may be necessary in certain situations.

5.12.7 Adequate ventilation shall be provided to ensure the atmosphere is maintained within acceptable limits.

5.13 Demolition

5.13.1 A detailed Method Statement detailing the demolition procedures/techniques to be used shall be submitted to and approved by the Engineer prior to commencement of work on site.

The Method Statement must include full details of measures to be taken to ensure that there are no persons remaining in the building/structure and to distance members of the public and Contractor's personnel from the building/structure prior to demolition.

5.14 Use of Explosives

5.14.1 The Contractor shall not use explosives without the written permission from the Engineer and relevant authorities (see Sub-Clause 2.5).

5.14.2 The Contractor shall observe all regulations regarding proper purchasing, transportation, storage, handling and use of explosives.

5.14.3 The Contractor shall ensure that explosives and detonators are stored in separate special buildings. These secured buildings shall be constructed, located and clearly marked in Arabic and English:

'DANGER—EXPLOSIVES'

as approved by the Engineer and relevant authorities (see Sub-Clause 2.5).

5.14.4 The Contractor shall ensure that all possible precautions are taken against accidental fire or explosion, and ensure that explosives and detonators are kept in a proper and safe condition.

5.14.5 The Contractor shall ensure that explosives and detonators are always transported in separate vehicles and kept apart until the last possible moment and that metallic tools are not used to open boxes of explosives or detonators.

5.14.6 Blasting Procedure: the Contractor shall carry out blasting operations in a manner that will not endanger the safety of persons and property. The Contractor shall, along with other necessary precautions:

(a) clear all persons from buildings and the area affected by the blasting. All such persons shall be given adequate notice of the actual time and date of blasting,

(b) ensure that police and other local authorities are kept fully informed, in advance, of the blasting programme so that they may be present when blasting takes place if they so require,

(c) erect warning notices around the area affected that blasting operations are in progress,

(d) carry out a thorough search of buildings and the area affected prior to blasting,
(e) ensure that blasting is only carried out by experienced shot firers. Priming, charging, stemming and shot firing shall be carried out with greatest regard for safety and in strict accordance with the rules and regulations of the relevant authorities (see Sub-Clause 2.5).
(f) ensure that explosive charges are not excessive, charged boreholes are properly protected and proper precautions are taken for the safety of persons and property,

5.14.7 The Contractor shall maintain an up-to-date inventory of all explosives and explosive devices and shall submit a monthly report to the Engineer, detailing the use of all explosives by date and location.

5.15 Excavation and Trenching

5.15.1 An excavation permit signed by the Engineer must be issued before excavation proceeds in any work location. The Contractor shall investigate and identify the location of existing services by study of the drawings, a visual/physical study of the site, sweeping by appropriate detection equipment and where necessary hand excavation of trial holes.

Following this investigation, the Contractor shall submit a written request for an excavation permit to the Engineer.

The Engineer will return the permit signed and dated to indicate:

—services which are to be maintained.
—services which are to be isolated.
—any special precautions to be taken.

A sample Excavation Permit is given in Annex 1.

5.15.2 The issue of an Excavation Permit by the Engineer shall not relieve the Contractor of his responsibilities under the Contract.

5.15.3 The side of all excavations and trenches exceeding 1.3 meters in depth which might expose personnel or facilities to danger resulting from shifting earth shall be protected by adequate temporary supports or sloped to the appropriate angle of repose.

5.15.4 All excavations, slopes and temporary supports shall be inspected daily and after each rain, before allowing personnel to enter the excavation.

5.15.5 Excavations 1.3 meters or more in depth and occupied by personnel shall be provided with ladders as a means for entrance and egress. Ladders shall extend not less than 1 meter above the top of the excavation.

5.15.6 The Contractor shall provide adequate barrier protection to all excavations. Barriers shall be readily visible by day or night.

5.15.7 Excavated or other materials shall not be stored at least 0.65 meters from the side of excavations.

5.15.8 The Contractor shall:

(a) ensure that stability and structural integrity of the Works are maintained during construction and shall provide temporary supports where necessary and shall not overload any part of the Works with materials, Plant or Contractor's Equipment.

(b) provide and maintain during the execution of the Works all shoring, strutting, needling and other supports as may be necessary to preserve stability of buildings, whether new or existing, on the site or adjoining property that may be endangered or affected by the Works.

(c) submit to the Engineer an outline of the methods proposed to be used for the support necessary to preserve stability of buildings or other structures, together with the relevant drawings, details, calculations, specifications and subsoil investigation, as necessary for approval. Such approval shall not pass to the Employer or the Engineer the responsibility for maintaining the stability of the buildings or relieve the Contractor from his responsibility.

5.16 Concrete Reinforcement Starter Bars

5.16.1 The Contractor shall ensure concrete reinforcement starter bars are not a danger to personnel. Where permitted by the Engineer, starter bars shall be bent down. Alternatively, the starter bars shall be protected using either hooked starters, plastic caps, plywood covers or other methods agreed with the Engineer.

6. *Environmental and health requirements*

6.1 Protection of the Environment

6.1.1 The Contractor shall be knowledgeable of and comply with all environmental laws, rules and regulations for materials, including hazardous substances or wastes under his control. The Contractor shall not dump, release or otherwise discharge or dispose of any such material without the authorization of the Engineer.

6.1.2 Any release of a hazardous substance to the environment, whether air, water or ground, must be reported to the Engineer immediately. When releases resulting from Contractor action occur, the Contractor shall take proper precautionary measures to counter any known environmental or health hazards associated with such release. These would include remedial procedures such as spill control and containment and notification of the proper authorities.

6.2 Air Pollution

6.2.1 The Contractor, depending on the type and quantity of materials being used, may be required to have an emergency episode plan for any releases to the atmosphere. The Contractor shall also be aware of local ordinances affecting air pollution.

6.2.2 The Contractor shall take all necessary measures to limit pollution from dust and any wind blown materials during the Works, including damping down with water on a regular basis during dry climatic conditions.

6.2.3 The Contractor shall ensure that all trucks leaving the Site are properly covered to prevent discharge of dust, rocks, sand, etc.

6.3 Water Pollution

6.3.1 The Contractor shall not dispose of waste solvents, petroleum products, toxic chemicals or solutions in the city drainage system or watercourse, and shall not dump or bury garbage on the Site. These types of waste shall be taken to an approved disposal facility regularly, and in accordance with requirements of relevant Authorities. The Contractor shall also be responsible to control all run-offs, erosion, etc.

6.4 Solid Waste

6.4.1 General Housekeeping

(a) The Contractor shall maintain the site and any ancillary areas used and occupied for performance of the Works in a clean, tidy and rubbish-free condition at all times.

(b) Upon the issue of any Taking-Over Certificate, the Contractor shall clear away and remove from the Works and the Site to which the Taking-Over Certificate relates, all Contractor's Equipment, surplus material, rubbish and Temporary Works of every kind, and leave the said Works and Site in a clean condition to the satisfaction of the Engineer. Provided that the Contractor shall be entitled to retain on Site, until the end of the Defects Liability Period, such materials, Contractor's Equipment and Temporary Works as are required by him for the purpose of fulfilling his obligations during the Defects Liability Period.

6.4.2 Rubbish Removal and Disposal

(a) The Contractor shall comply with statutory and municipal regulations and requirements for the disposal of rubbish and waste.
(b) The Contractor shall provide suitable metal containers for the temporary storage of waste.
(c) The Contractor shall remove rubbish containers from site as soon as they are full. Rubbish containers shall not be allowed to overflow.
(d) The Contractor shall provide hardstandings for and clear vehicle access to rubbish containers.
(e) The Contractor shall provide enclosed chutes of wood or metal where materials are dropped more than 7 meters. The area onto which the material is dropped shall be provided with suitable enclosed protection barriers and warning signs of the hazard of falling materials. Waste materials shall not be removed from the lower area until handling of materials above has ceased.
(f) Domestic and biodegradable waste from offices, canteens and welfare facilities shall be removed daily from the site.
(g) Toxic and hazardous waste shall be collected separately and be disposed of in accordance with current regulations.
(h) No waste shall be burnt on site unless approved by the Engineer.

6.4.3 Asbestos Handling and Removal

The Contractor shall comply with all local regulations regarding the handling of asbestos materials. In the absence of local regulations, relevant International Standards shall apply.

6.4.4 Pest Control

The Contractor shall be responsible for rodent and pest control on the Site. If requested, the Contractor shall submit to the Engineer, for approval, a detailed programme of the measures to be taken for the control and eradication of rodents and pests.

6.5 Noise Control

6.5.1 The Contractor shall ensure that the work is conducted in a manner so as to comply with all restrictions of the Authorities having jurisdiction, as they relate to noise.

6.5.2 The Contractor shall, in all cases, adopt the best practicable means of minimizing noise. For any particular job, the quietest available plant/and or machinery shall be used. All equipment shall be maintained in good mechanical order and fitted with the appropriate silencers, mufflers or acoustic covers where applicable. Stationary noise sources shall be sited as far away as possible from noise-sensitive areas, and where necessary acoustic barriers shall be used to shield them. Such barriers may be proprietary types, or may consist of site materials such as bricks or earth mounds as appropriate.

6.5.3 Compressors, percussion tools and vehicles shall be fitted with effective silencers of a type recommended by the manufacturers of the equipment. Pneumatic drills and other noisy appliances shall not be used during days of rest or after normal working hours without the consent of the Engineer.

6.5.4 Areas where noise levels exceed 90 decibels, even on a temporary basis, shall be posted as high noise level areas.

6.6 Protection of Archaeological and Historical Sites

6.6.1 Excavation in sites of known archaeological interest should be avoided. Where this is unavoidable, prior discussions must be held with the Directorate of Antiquities in order to undertake pre-construction excavation or assign an archaeologist to log discoveries as construction proceeds. Where historical remains, antiquity or any other object of cultural or archaeological importance are unexpectedly discovered during construction in an area not previously known for its archaeological interest, the following procedures should be applied:

a) Stop construction activities.
b) Delineate the discovered site area.
c) Secure the site to prevent any damage or loss of removable objects. In case of removable antiquities or sensitive remains, a night guard should be present until the responsible authority takes over.
d) Notify the responsible foreman/archaeologist. Who in turn should notify the responsible authorities, the General Directorate of Antiquities and local authorities (within less than 24 hours).
e) Responsible authorities would be in charge of protecting and preserving the site before deciding on the proper procedures to be carried out.
f) An evaluation of the finding will be performed by the General Directorate of Antiquities. The significance and importance of the findings will be assessed according to various criteria relevant to cultural heritage including aesthetic, historic, scientific or research, social and economic values.
g) Decision on how to handle the finding will be reached based on the above assessment and could include

changes in the project layout (in case of finding an irrevocable remain of cultural or archaeological importance), conservation, preservation, restoration or salvage.

h) Implementation of the authority decision concerning the management of the finding.

i) Construction work could resume only when permission is given from the General Directorate of Antiquities after the decision concerning the safeguard of the heritage is fully executed.

6.6.2 In case of delay incurred in direct relation to Archaeological findings not stipulated in the contract (and affecting the overall schedule of works), the contractor may apply for an extension of time. However the contractor will not be entitled for any kind of compensation or claim other than what is directly related to the execution of the archaeological findings works and protections.

7. *Additional requirements for work in public areas*

7.1 General

7.1.1 These additional requirements shall apply to all works carried out in Public Areas.

7.1.2 Public Areas are defined as areas still used by or accessible to the public. These include public roads and pavements, occupied buildings and areas outside the Contractor's boundary fencing.

7.1.3 All work in Public Areas shall be carried out to minimize disturbance and avoid dangers to the public.

7.1.4 Before commencing work, the Contractor shall ensure that all necessary resources, including labour, plant and materials, will be available when required and that the works will proceed without delays and be completed in the shortest possible time. Periods of inactivity and slow progress or delays in meeting the agreed programme for the works, resulting from the Contractor's failure to provide necessary resources or other causes within the control of the Contractor, will not be accepted. In the event of such inactivity, slow progress or delays, the Contractor shall take immediate action to rectify the situation, including all possible acceleration measures to complete the works within the agreed programme. Details of the actions and acceleration measures shall be submitted to the Engineer. If the Engineer is dissatisfied with the Contractor's proposals, the Contractor shall take such further actions or measures as required by the Engineer. All costs incurred shall be the responsibility of the Contractor.

7.2 Method Statement

7.2.1 The Contractor shall submit to the Engineer a method statement for each separate area of work in Public Areas. The Method Statement shall include:

- a general description of the Works and methodology of how it will be carried out.
- details of the measures and temporary works to minimize disturbance and safeguard the public. These shall include temporary diversions, safety barriers, screens, signs, lighting, watchmen and arrangements for control of traffic and pedestrians, and an advance warning to be given to the public.
- details of temporary reinstatement and maintenance of same prior to final reinstatement.
- for works involving long lengths of trenches or works to be completed in sections, the lengths or sections of each activity (e.g., up to temporary reinstatement, temporary reinstatement, final reinstatement) to be carried out at any one time.
- details of the availability of necessary resources (labour, plant, materials, etc.) to complete the work.
- a programme showing start and completion dates and periods for all activities of each length or section, including temporary works, and the works overall.
- such further information as necessary or required by the Engineer.

7.2.2 The Contractor shall not commence work, including temporary works, until approval of the Contractor's Method Statement by the Engineer.

7.2.3 Method Statements shall be updated based on actual progress or as and when required by the Engineer.

7.3 Closure of Roads, etc.

7.3.1 The closure or partial closure of roads, pavements and other public areas will only be permitted if approved by the Engineer and Relevant Authorities. The Contractor shall detail for each closure the extent of area to be closed, the reasons and duration of the closure and, where appropriate, proposed diversions.

A sample Street Closure Permit is given in Annex 2.

7.3.2 Access to Properties Affected by the Works:

The Contractor shall identify, protect and maintain accesses to all properties affected by the works.

7.3.3 The Contractor shall ascertain and comply with any regulations concerning traffic and parking in addition to the obligations imposed by the Conditions of Contract and by law.

7.3.4 The Contractor shall provide and maintain all necessary diversion, diversion signs, barricades, fencing, lighting, flagmen or slop/go Signs where the Works affect the safely of traffic and the public on existing roads or temporary diversion roads.

7.4 Trench and Other Excavations

7.4.1 The requirements covering trench and other excavations will depend on the location and type of the excavation and the potential risks to the public.

7.4.2 The following guidelines apply particularly to trenches but shall also apply to other types of excavations:

(a) before commencing work the Contractor shall:

- notify the Engineer on the location and duration of the work. An excavation permit signed by the Engineer must be issued in accordance with Sub-Clause 5.15.1 before excavation proceeds in any work location;
- obtain permission from relevant authorities including the police when required. The Contractor's attention is drawn to the requirements of Legislative Decree No 68 dated 9 September 1983, issued by the President of the Republic of Lebanon, and in particular to the provisions therein regarding prior notification by the Contractor to and the issue of excavation licenses by the Director of Roads or the Head of the Municipal Authority concerned, as applicable, before the commencement of excavations within the limits of streets, roads and other areas defined under the said Decree;
- erect all temporary works such as barriers, warning signs, lighting, etc;
- have available adequate materials for temporary supports to sides of excavations and necessary labour, plant and materials to complete the work within the shortest possible time.

(b) in carrying out the works the Contractor shall, unless otherwise permitted or required by the Engineer:

- not open more than one excavation within a radius of 250 metres;
- limit the length of trench excavation open at one time to 150 metres;
- maintain and alter or adapt all temporary works including supports to sides of excavations;

- remove all surplus excavated material the same day it is excavated;
- complete the works, including final reinstatement within ten days;
- where final reinstatement is not achieved within the required time, to carry out temporary reinstatement;
- ensure that any temporary reinstatement is maintained at the correct level until final reinstatement is achieved.

7.4.3 The above guidelines shall not relieve the Contractor of his obligations and responsibilities.

7.5 Safety Barriers

7.5.1 Safety barriers shall be provided to the perimeter of work areas and to trench and other types of excavations and to existing openings such as manholes, draw pits and the like. When exposed to the public, safety barriers shall be provided to both sides of trenches and around all sides of openings.

7.5.2 The Contractor shall provide details of the type or types of safety barriers for each excavation for the approval of the Engineer prior to commencing work. No work shall commence until the safety barriers are in place.

7.5.3 The type of safety barrier used shall be appropriate to the particular location and the potential risks to the public. Examples of different types of safety barriers are given below along with attached figures:

- Type 1—excavated material;
- Type 2—non-rigid barrier of rope or florescent tape strung between metal rods driven into the ground;
- Type 3—non-rigid barrier type K2, K5a, K5c and K8.
- Type 4—rigid concrete barrier. Such barriers should be secured by means of dowels driven into the ground.

7.5.4 The following are guidelines on the type of safety barriers that could be used in differing situations. They apply particularly to trenches but also apply to other types of excavations, existing openings and to the perimeter of work areas:

- areas not subject to vehicular traffic—Types 1 or 2;
- roadways (low traffic speed)—Types 1 and 3 or Types 2 and 3;
- roadways (high traffic speed)—Type 3 (short term 1 to 2 days) or Type 4 (long term more than 2 days).

7.5.5 The above examples of the types of barriers and the guidelines on situations in which they could be used shall not relieve the Contractor of his obligations and responsibilities.

8. Contractor's site check list

8.1 A sample Contractor's Site Check List is included in Annex 3. This is included to assist contractors should they wish to introduce such a system as part of their site management procedures. The list is not exhaustive and further items will need to be added by the Contractor.

8.2 The list is issued for guidance only, and does not, in any way, revise or limit the requirements covered elsewhere in these Regulations.

9. Protection of other property and services

9.1 Roads and Footpaths: the Contractor shall protect public and private roads, footpaths and the like from damage by site traffic or other causes arising from the execution of the Works and shall repair any damage to the satisfaction of the relevant public authority or private owner.

9.2 Trees, Hedges, Shrubs, Lawns: the Contractor shall protect and preserve, trees, hedges, shrubs, lawns etc., and shall replace to approval, or treat as instructed, any plants or areas damaged or removed without approval.

9.3 Existing Features: the Contractor shall prevent damage to existing buildings, fences, gates, walls, roads, paved areas and other features on the Site or adjacent thereto which are to remain in position during the execution of the Works.

9.4 Existing Services

The Contractor shall:

notify all service authorities and private owners before commencing any work which may affect or damage existing drains and services and observe all service authorities' regulations and/or recommendations work adjacent to existing services.

ascertain the positions of all services not indicated in the Contract Documents and check the positions of those which are so indicated.

adequately protect, maintain and prevent damage to all services and shall not interfere with their operation without the consent of the service authority or owner.

If any damage is caused to existing services as a result of execution of the Works, the Contractor shall notify the Project Manager Representative/ Engineer's Representative and the service authority or private owner and make arrangements to repair the damage to the satisfaction of the service authority or private owner as appropriate.

9.5 Adjoining Property
The Contractor shall:
- take all reasonable precautions to prevent damage to adjoining property and, if any damage is caused as a result of the execution of the Works, make good to the satisfaction of the owner.
- obtain permission of the owners if it is necessary to erect Temporary Works or otherwise use adjoining property and pay all charges.
- advise owners or occupiers of adjoining property of the dates on which work, which may affect them, is to be executed

9.6 Existing Condition of roads, paths, features, services and adjoining property which is at risk from damage shall be recorded by photographs or surveys as appropriate.

9.7 Occupied Premises
The Contractor shall:
(a) where the works are to be carried out in or around occupied premises ascertain the times and nature of the occupation and use. Carry out the Works with minimum inconvenience, nuisance and danger to the occupants and users.
(b) if the danger to the occupied premises is such as to involve the safety of persons advise the Employer to evacuate temporarily such persons until the danger is eliminated. The expense of evacuation, temporary accommodation and re-occupation of the premises and other expenses shall be borne by the Employer.

Annex 1

Sample Excavation Permit

To: .... (Engineer)

From: .... (Contractor)
....Date: ....

CDR Contract No: ....
Request for Excavation Permit No: ....

Please give approval for excavation to proceed in the following area:

Work to start on:

Existing services have been checked and identified by:

Drawings	#	Physical Survey	#
CAT scan	#	Trial Holes Excavation	#

Signed (Contractor): ....

Approval of Engineer

The above excavation may proceed, subject to the following:

Services to be maintained:

Services to be isolated before work proceeds:

Other matters:

Signed (Engineer): ....

Date: ....

Annex 2

Sample Street Closure Permit

To: .... (Engineer)

From: .... (Contractor)
....Date: ....

CDR Contract No: ....
Request for Street Closure Permit No: ....

Please give approval for the closure of the following street(s) from ....to ....
(dates)

Street(s):

Reasons:

Proposed diversions:

Signed (Contractor): ....

Approval of the Engineer

The above street(s) may be closed for the periods stated subject to the following conditions:

Approval has been given by relevant authorities and the police;

Other:

Signed (Engineer): ....

Date: ....

Annex 3

Sample Contractor's Site Check List

Safe Access:

- arrangements for visitors and new workers to the Site.
- safe access to working locations.
- walkways free from obstructions.
- edge protection to walkways over 2m above ground.
- holes fenced or protected with fixed covers.
- tidy Site and safe storage of materials.
- waste collection and disposal.
- chutes for waste disposal, where applicable.
- removal or hammering down of nails in timber.
- safe lighting for dark or poor light conditions.
- props or shores in place to secure structures, where applicable.

Ladders:

- to be used only if appropriate.
- good condition and properly positioned.
- located on firm, level ground.
- secure near top. If not possible, to be secured near the bottom, weighted or footed to prevent slipping.
- top of ladder minimum 1 metre above landing place.

Scaffolding:

- design calculations submitted.
- proper access to scaffold platform.
- properly founded uprights with base plates.
- secured to the building with strong ties to prevent collapse.
- braced for stability.
- load-bearing fittings, where required.
- uprights, ledgers, braces and struts not to be removed during use.
- fully boarded working platforms, free from defects and arranged to avoid tipping or tripping.
- securely fixed boards against strong winds.
- adequate guard rails and toe boards where scaffold 2 metres above ground.
- designed for loading with materials, where appropriate.
- evenly distributed materials.
- barriers or warning notices for incomplete scaffold (i.e. not fully boarded).
- weekly inspections and after bad weather by competent person.
- record of inspections.

Excavation:

- underground services to be located and marked and precautions taken to avoid them.
- adequate and suitable timber, trench sheets, props and other supporting materials available on Site before excavation starts.
- safe method for erecting/removal of timber supports.
- sloped or battered sides to prevent collapse.
- daily inspections after use of explosives or after unexpected falls of materials.
- safe access to excavations (e.g. sufficiently long ladder).
- barriers to restrict personnel/plant.
- stability of neighboring buildings risk of flooding.
- materials stacked, spoil and vehicles away from top of excavations to avoid collapse.
- secured stop blocks for vehicles tipping into excavations.

Roof work:

- crawling ladders or boards on roofs more than 10 degrees.
- if applicable, roof battens to provide a safe handhold and foothold.
- barriers or other edge protection.
- crawling boards for working on fragile roof materials such as asbestos cement sheets or glass.
- Guard rails and notices to same.

- roof lights properly covered or provided with barriers.
- during sheeting operations, precautions to stop people falling from edge of sheet.
- precautions to stop debris falling onto others working under the roof work.

Transport and mobile plant:

- in good repair (e.g. steering, handbrake, footbrake).
- trained drivers and operators and safe use of plant.
- secured loads on vehicles.
- passengers prohibited from riding in dangerous positions.
- propping raised bodies of tipping lorries prior to inspections.
- control of on-site movements to avoid danger to pedestrians, etc.
- control of reversing vehicles by properly trained banksmen (a banksman is the skilled person who directs the operation of a crane or larger vehicle from the point near where loads are attached and detached), following safe system of work.

Machinery and equipment:

- adequate and secured guards in good repair to dangerous parts, e.g. exposed gears, chain drives, projecting engine shafts.

Cranes and lifting appliances:

- weekly recorded inspections.
- regular inspections by a competent persons.
- test certificates.
- competent and trained drivers over 18 years of age.
- clearly marked controls.
- checks by driver and banksman on weight ofload before lifting.
- efficient automatic safe load indicator, inspected weekly, for jib cranes with a capacity of more than one tone.
- firm level base for cranes.
- sufficient space for safe operation.
- trained banksman/slinger to give signals and to attach loads correctly, with knowledge of lifting limitations of crane.
- for cranes with varying operating radius, clearly marked safe working loads and corresponding radii.
- regular maintenance.
- lifting gear in good condition and regularly examined.

Electricity:

- measures to protect portable electric tools and equipment from mechanical damage and wet conditions.
- checks for damage to or interference with equipment, wires and cables.
- use of the correct plugs to connect to power points.
- proper connections to plugs; firm cable grips to prevent earth wire from pulling out.
- "permit-to-work" procedures, to ensure safety.
- disconnection of supplies to overhead lines or other precautions where cranes, tipper lorries, scaffolding, etc. might touch lines or cause arcing.

Cartridge operated tools:

- maker's instruction being followed.
- properly trained operators, awareness of dangers and ability to deal with misfires.
- safety goggles.
- regular cleaning of gun.
- secure place for gun and cartridges when not in use.

Falsework/formwork:

- design calculations submitted.
- method statement dealing with preventing falls of workers.
- appointment offalsework coordinator.
- checks on design and the supports for shuttering and formwork.
- safe erection from steps or proper platforms.
- adequate bases and ground conditions for loads.
- plump props, on level bases and properly set out.
- correct pins used in the props.
- timberwork in good condition.
- inspection by competent person, against agreed design before pouring concrete.

Risks to the Public:

- identify all risks to members of the public on and off Site, e.g. materials falling from scaffold, etc., Site plant and transport (access/egress) and implement precautions, e.g. scaffold fans/nets, banksmen, warning notices, etc..
- barriers to protect/isolate persons and vehicles.
- adequate site perimeter fencing to keep out the public and patticularly children. Secure the Site during non-working periods.
- make safe specific dangers on site during non-working periods, e.g. excavations and openings covered or fenced, materials safely stacked, plant immobilised, ladders removed or boarded.

Fire - general:

- sufficient number and types of fire extinguishers.
- adequate escape routes, kept clear.
- worker awareness of what to do in an emergency.

Fire - flammable liquids:

- proper storage area.
- amount of flammable liquid on Site kept to a minimum for the day's work.
- smoking prohibited; other ignition sources kept away from flammable liquids.
- proper safety containers.

Fire - compressed gases, e.g. oxygen, LPG, acetylene:

- properly stored cylinders.
- valves fully closed on cylinders when not in use.
- adopt "hot work" procedures.
- Site cylinders in use outside huts.

Fire - other combustible materials:

- minimum amount kept on Site.
- proper waste bins.
- regular removal of waste material.

Noise:

- assessment of noise risks.
- noisy plant and machinery fitted with silencers/muffs.
- ear protection for workers if they work in very noisy sunoundings.

Health:

- identify hazardous substances, e.g. asbestos, lead, solvents, etc. and assess the risks.
- use of safer substances where possible.
- control exposure by means other than by using protective equipment.
- safety information sheets available from the supplier.
- safety equipment and instructions for use.
- keep other workers who are not protected out of danger areas.
- testing of atmosphere in confined spaces; provision of fresh air supply if necessary. Emergency procedures for rescue from confined spaces.

Manual handling:

- avoid where risk of injury.
- if unavoidable, assess and reduce risks.

Protective clothing:

- suitable equipment to protect the head, eyes, hands and feet where appropriate.
- enforce wearing of protective equipment.

Welfare:

- suitable toilets.
- clean wash basin, hot/warm water, soap and towel.
- room or area where clothes can be dried.
- wet weather gear for those working in wet conditions.
- heated site hut where workers can take shelter and have meals with the facility for boiling water.
- suitable first aid facilities.

Work in Public Areas

- all risks to the public identified.
- method statement approved.
- road closures approved.
- temporary diversions in place.
- safety barriers erected/maintained.
- safety signs and lighting installed/maintained.
- labour, materials, plant and other resources sufficient to meet programme.
- temporary reinstatement completed and properly maintained.
- permanent reinstatement completed at earliest possible date.

chapter nine

Fostering a strong positive safety climate with contractors

Tristan Casey
Sentis

Autumn D. Krauss, PhD
Sentis

Contents

Given the extent to which organisations in heavy industries rely on contractors to accomplish critical work tasks while maintaining safe operations, it is important to identify ways in which contractors can be better integrated into the workforce with the goal of improving not only their own safety performance but also the safety outcomes for the entire operation. To accomplish this, we must consider how well-researched and understood concepts such as *safety climate* might function differently when contractors (either contractor organisations or individual contractor employees) are involved. With this objective, this chapter describes the unique characteristics of the organisational social context that contractors work within along with the particular challenges associated with integrating contractors into a client organisation and its established workgroups. The safety implications of this contractor work reality are explored, with strategies to achieve exemplary contractor safety management offered within the context of psychological best practices for workplace safety.

A contractor's workplace safety reality

Across industries, companies are increasingly calling upon contractor organisations or individual contract-based workers to perform work in hazardous environments (e.g., offshore oil and gas installations). Indeed, many organisations rely on these contractors to achieve business-critical outcomes. Contractor organizations and their workers provide highly specialised skills and knowledge, facilitate swift expansion into new frontiers, operate and maintain critical infrastructure such as plants and equipment, and reduce the burden of a large permanent workforce. Even with serving these important functions, the strategic and purposeful social integration of contractors into client organisations is often mismanaged at best or at worst absent (Boyce, Ryan, Imus and Morgeson, 2007). The tangible consequences of ineffective contractor social integration often include reduced operational efficiency and productivity not to mention increased safety risks and hazards.

The ultimate goal is for safety to be at least maintained if not improved as organisations incorporate contractors into their workforce. To accomplish this, contractors must be successfully integrated into the client organisation's existing social structure. Admittedly, this is not an easy task. Contractor organizations vary markedly in the maturity of their safety approaches, with their employees commonly being drawn from diverse cultural backgrounds, feeling pressure from themselves or others to work longer and faster, and being required to negotiate complex and confusing supervisory structures (Clarke, 2003; Lingard, Cooke and Blismas, 2010). Contractors are also often placed under considerable financial and competitive pressure, which can result in an enhanced emphasis on production at the expense of safety. Finally, contractors often exist on the periphery of client organisations with reduced support and communication as compared to the permanent workforce. Overall, the nature of social interaction and group dynamics increases the interpersonal challenges faced by contractor organizations and their workers. These social factors—if left unmanaged or managed ineffectively—can drastically decrease the safety of this work population, not to mention the safety of operations in general.

In light of the pressures outlined above, it is not surprising that contractors typically experience worse safety outcomes than permanent employees (Clarke, 2003; Park and Butler, 2001). These poorer safety outcomes in part result from contractors being more likely to engage in unsafe work practices, which significantly increases their risk of workplace injury. More specifically, contract workers are more likely than permanent workers to comply with ineffective or 'bad' safety procedures, likely due to fear of job loss or social/financial penalties, and apply effective or 'good' safety procedures at the wrong time, likely due to lack of safety and/or task-specific knowledge or experience. Further, contractors may have increased difficulty operating in ambiguous safety situations and be more prone to make inaccurate hazard identifications and risk assessments.

In summary, research suggests that contractors are an at-risk population with unique workplace challenges and poorer safety outcomes than permanent. Moreover, a contributing factor to this contractor experience is the difficulty with integrating contractors into the client organisation's social structure and dynamics. Within the safety context, safety climate is the fundamental concept that describes the social underpinnings of employees' safety attitudes, behaviours and results. Given the substantial evidence indicating the large effect of a company's safety climate on its safety performance (Clarke, 2006), we suggest that further examination of the impact of contractors on an organisation's safety climate is warranted. Likewise, targeted safety climate interventions may be effective in improving workplace safety when contractors are present.

Safety climate as a psychological lever to improve workplace safety

An organisation is a complex structure of people, practices and resources that must work cohesively in order to survive and prosper. Further, an organisation establishes core objectives that employees work to accomplish in exchange for benefits such as pay, training and development, along with a sense of purpose and belonging. Accordingly, employees must discern what is valued by their organisation in order to develop a shared understanding of daily life on-the-job and achieve these objectives efficiently and harmoniously. In other words, employees seek to understand what behaviours are rewarded, what behaviours are punished, and which tasks and work practices should be prioritised. To get answers to these questions, employees look for signals based on formal policies and procedures espoused by their organisation as well as cues that are more implicit and produced by informal practices adopted by their workgroup (Schneider and Reichers, 1983).

Messages from senior management, official documentation, induction processes, data from information systems, and rules, guidelines, and procedures all convey information about what types of behaviour are expected from employees. At the highest level, executives and senior managers chart the course of the organisation by establishing strategic goals, developing policy statements and allocating resources to initiate action in certain areas of the business. These formal and explicit sources of information provide workers with a high-level understanding of how to behave within their broader organisation's social and operational contexts.

That being said, employees perform daily tasks within the more immediate context of their workgroup. Through their words and actions, members of the workgroup offer insight about 'how things *really* work,' which may be in direct conflict with what is formally espoused by management. Within the workgroup, formal organisational policies and procedures are interpreted and implemented, and the extent to which this implementation is aligned with management's directive exerts a powerful influence over employees' thinking and behaviour. For example, workgroup supervisors exercise some level of discretion in how formal procedures are put into practice and must also interpret senior management's vision and priorities to direct workers towards the achievement of organisational goals. Consequently, employees can form very different perceptions of the organisation's stance on workplace issues such as safety according to the degree of (mis)alignment between management espousals and supervisor enactments across levels of the organisational hierarchy (Zohar, 2010).

Perceptions of these formal/espoused and informal/enacted policies, practices, and procedures create what is known as a *psychological climate* (Rousseau, 1988), and where these perceptions are shared with other employees, an *organisational or team climate* (Ashforth, 1985) is said to exist. Within a workgroup, shared perceptions are typically created by members engaging in a sense-making process of looking for patterns among what is espoused by management and enacted by the group. Within organisations engaged in safety-critical operations, workers seek to understand what is expected of them through visible signs of the priority of safety compared to production, leadership behaviours such as praise and punishment for particular safety actions, and the nature of co-worker safety interactions (Zohar, 1980; Zohar, 2010). It is through these shared employee perceptions about the value placed on safety by their organisation that a safety climate is created.

This safety climate is a momentary snapshot of the underlying safety culture, essentially the manifestation of the less readily observed aspects of an organisation's safety values (Zohar, 2010). Also, whereas safety culture is comprised of individual employees' safety beliefs and values that are well ingrained and hence fairly stable over time, safety climate is more dynamic and responsive to changes implemented by the organisation (e.g., the introduction of a new safety procedure or safety recognition program). Accordingly, climate and culture are intimately related; over time, safety climate can influence the underlying safety culture as employees' shared perceptions about their organisation's prioritisation of safety can begin to alter employees' individual attitudes and beliefs about safety. Therefore, organisations should strive to achieve a favourable safety climate and use it as a lever to move the underlying safety culture in a positive direction.

What value an organisation places on safety and how this value is reflected by supervisor and co-worker behaviour determines a safety climate's level and strength. Whereas *safety climate level* refers to the positivity or negativity of employees' perceptions about the company's prioritisation of safety, *safety climate strength* refers to the consistency or agreement of perceptions across employees (Schulte, Ostroff, Shmulyian and Kinicki, 2009).

An organisation's key safety objective should be to build a strong positive safety climate, because it is associated with effective safety behaviour among employees, and hence positive organisational safety outcomes. A positive safety climate directly influences employees' safety behaviour by providing cues indicating behaviours such as compliance with safety procedures and helping others to work safely are valued and rewarded by the organisation (Clarke, 2006); where the safety climate is also strong, these positive safety behaviours will be more consistently demonstrated across the organisation (Zohar and Luria, 2004). Unfortunately, organisations that utilise contractors within their operations are likely to encounter challenges with establishing a strong positive safety climate for reasons that range from cross-cultural obstacles to issues with the transient nature of a contractor workforce.

Adding contractor complexity to safety climate

Simply put, contractors add a layer of complexity to the process of achieving a strong positive safety climate. Contract workers may either identify with an external contractor employer or as a sole-entity in the case of self-employed contractors, both of which impact the client organisation's safety climate and, subsequently, its safety performance. In the former case, research supports the notion that contractors differentiate between the safety climates of their own organisation and their contracting client organisation (Lingard et al., 2010). Specifically, research conducted in the Australian construction industry demonstrated that the principal contractor's safety climate (at both organisational and team levels) contributed indirectly to employees' injuries via its influence on employees' perceptions of their own organisation's safety climate (the subcontracted organisation). Therefore, contractor safety management depends on the critical interactions that occur between the safety climates of all organisations engaged to work on the same project or site.

In the latter case, contractors who identify as individuals (e.g., self-employed or new to the industry) can experience difficulty integrating within organisations, and hence are less influenced by the client's safety climate even if it could be a source of positive influence. In this situation, contractors consider themselves as independent and somewhat temporary operators, likely focussed more on transactional outcomes such as pay and benefits rather than social interactions with leaders and co-workers. As a result, these workers may experience low job satisfaction and organisational commitment as well as disconnection from communication and social networks (Clarke, 2003). Indeed, research with temporary workers has shown that they tend to rely on personal knowledge and skills rather than social cues such as safety climate to understand their place in the organisation and inform their behaviour (Rogers, 1995; Luria and Yagil, and Luria, 2010). Further, given that temporary workers are less likely to receive training and development opportunities, even notably in safety, and have much shorter tenure (and hence, less workplace-specific experience) within a particular organisation, it is not surprising that this group experiences a higher rate of workplace injury than permanent employees.

In sum, much research has demonstrated that employees engage in sense-making to understand what safety behaviours are rewarded and valued by organisations. Communications from senior management about safety priorities and policies provide one source of information for this purpose. Within teams, supervisors' interpretations of safety procedures and their specific safety leadership behaviours (e.g., reward and recognition) as well as the safety practices of co-workers provide additional salient sources of data (Zohar, 2010). Where employees' perceptions of these organisational- and team-level safety factors are shared with co-workers, a safety climate exists. This safety climate may vary in both level (positivity) and strength (consistency), which has marked effects on employee safety behaviour and injury outcomes.

Contractors are particularly challenged when integrating into the established work structure and social dynamics of an organisation, which includes the organisation's safety climate. Importantly, the ultimate goal of establishing a strong positive safety climate can be impeded if the social aspects of contractor safety are ignored or mismanaged. The particular social issues that must be effectively managed by organisations employing contractors include the following: transient/temporary employment, national culture, discrepancies between safety climates, and differential treatment. Each of these contractor social challenges is summarised in the text following, with particular reference to its impact on safety climate.

Transient/temporary employment. Contractors are typically employed on a temporary basis, such as for the duration of a project or during a period of high work demand (e.g., production shutdown for maintenance). Although employing contractors often enables organisations to achieve operational goals on time and within budget, there can be difficulties with maintaining a strong safety climate across the worksite or organisation when contractors are involved. This is because organisational tenure is related to safety climate strength—as individuals spend time working within organisations and 'learning the ropes,' their safety climate perceptions become progressively homogenous.

According to organisational psychology, two processes explain this effect: attraction–selection–attrition (ASA; Schneider, Goldstein and Smith, 1995) and socialisation (Lindell and Brandt, 2000). From the ASA perspective, individuals with like-minded attitudes, preferences, and behavioural tendencies are attracted to, selected to, and retained within a particular organisation; whereas those with dissimilar characteristics are typically not interested in applying to, not hired to, or not keen to stay working at that organisation. In addition, the process of organisational socialisation suggests that through daily interactions with co-workers, new employees develop more accurate perceptions of, and knowledge about, the inner workings of their organisation (Allen, 2006). Consequently, as employee tenure increases, the volume of these social interactions increases, enhancing the consistency of employees' climate perceptions.

Research clearly supports the application of these psychological theories to contractor safety management. First, organisational tenure has been shown to directly predict safety climate strength, such that longer average employee tenure within a worksite is associated with stronger (more consistent) worksite safety climate (Beus, Bergman and Payne, 2010). Second, safety climate strength is positively impacted by more frequent and better-quality workgroup interactions, with higher social network density predicting increased safety climate strength among army platoons (Zohar and Tenne-Gazit, 2008). Thus, given that contractors are more likely to have shorter organisational tenure than permanent employees and be less likely to interact socially with other workers, the development of a strong safety climate within an organisation employing a large contractor population or bringing on many contractors for a specific project may be particularly challenging.

National culture. As companies in heavy industry expand operations overseas, increasing numbers of migrant workers employed through contractors are working under the management of, or alongside, Western employees. In this case, workgroups are often comprised of workers from diverse ethnic, cultural, and ideological backgrounds. Indeed, multicultural integration has long been an issue for companies in the Middle East and Asia who frequently employ contractors from Africa, Eastern Europe, and India. Still, migrant contractors experience a higher rate of injury as compared to their Western counterparts (Starren, Hanrikx and Luijters, 2013), which points to issues with contractor safety management within a multicultural setting. Case in point: national culture is rarely considered in the design and implementation of safety procedures or training, apart from possible translation of information into other languages (Vickers, Baldock, Smallbone, James and Ekanem, 2003).

Moreover, national culture may directly impact workers' safety performance through the interplay of different worldviews or perspectives on safety-relevant issues. As an example, national culture has been shown to impact risk perception accuracy, safety climate perceptions, and other safety-relevant assumptions and beliefs such as obedience to authority (Starren et al., 2013). Of particular importance to contractor safety management, several studies have shown that perceptions of risk differ across cultures (e.g., Kouabenan, 2009; Renn and Rohrmann, 2000). There is also some evidence that workers from Asian countries tend to be more optimistic about risk as compared to Western workers; however, there are likely even complexities within the Western worker population, given subtle differences between European and American workers in individualism and collectivism. Even attempts to improve risk perception through worker safety training can be hampered by national culture. Specifically, low tolerance for ambiguity, which is a dimension of national culture (Hofstede, 1993), has been shown to dilute some of the safety training benefits for preventing workplace injury (Burke, Chan-Serafin, Salvador, Smith and Sarpy, 2008). Workers from cultures with low tolerance for ambiguity and uncertainty are less likely to effectively manage hazards in ambiguous work situations even after experiencing safety training.

Furthermore, cultural differences may hinder organisational efforts to establish a strong positive safety climate, primarily because of increased likelihood of work conflict, discrimination and poor communication within culturally diverse workgroups (Starren et al., 2012). In organisations where management of cross-cultural and diversity issues is poor, workers have fewer opportunities to communicate openly and effectively, which is essential for the development of shared perceptions about safety.

Finally, migrant workers' safety performance may be further impeded by underlying beliefs and assumptions that vary with national culture. According to Hofstede (1993), cultures differ along dimensions, including power distance (acceptance of power inequality), masculinity/femininity (concern for social relationships), and individualism/collectivism (identification with the group). Clearly, workers' safety beliefs and behaviour would be influenced by all of these cultural dimensions, with 'blind' obedience or compliance, concern for the entire team's safety and well-being, and level of safety communication all varying according to cultural background. Overall, these findings illuminate the complexities of safety management when contractors include employees from different cultural backgrounds.

Safety climate discrepancies. As discussed earlier, an organisation develops a safety climate reflective of its underlying safety culture, and the extent to which workers' perceptions are favourable and consistent determines the level and strength, respectively, of the safety climate. In part, employees' perceptions of their organisation's safety climate are driven by the maturity and sophistication of its safety management strategy, including associated systems and practices. Contractor organisations are presented with many safety management challenges, including limited resources to invest in safety, production pressure from the principal contractor and/or client organisation, and competition from other contractors. These challenges have the potential to reduce a contractor organisation's focus on safety and consequently the positivity of its safety climate and the effectiveness of its safety performance.

Remember, contract workers differentiate between the safety climate of their own organisation and that of the client organisation, with these two safety climates influencing each other in meaningful ways (Lingard et al., 2010). Relatedly, safety climate operates at the organisational and workgroup levels, with organisational and workgroup safety climates potentially differing in level and strength. That being said, there is usually still some alignment between organisational and workgroup safety climates, because they are both being influenced by the same safety policies and supervisors do have limits on their discretion in enacting the policies. In the case of the contractor and client organisations, there is no such guarantee of alignment, given the possibility of completely independent approaches to safety management.

Discrepancies in safety climates between levels of organisational hierarchy or organisations themselves have important implications for safety performance. Research suggests that regardless of senior management's commitment to safety, workers will pay more attention to team-level cues when forming their perceptions of safety climate; this is particularly the case when there are discrepancies in safety values between levels of management (Zohar, 2010). The corollary of this finding is that when the contractor organisation's safety climate is less positive than that of the client organisation, the contract workers' safety climate perceptions are likely to be overall less favourable and their safety performance is likely to be lower. Taken together, client organisations should not only strive to create a strong positive safety climate within their own organisations but also seek to support the efforts of contractor organisations as they develop their own safety climates.

Differential treatment. It is human nature to form social groups based on observed or perceived characteristics of other people. According to social identity theory, classification assists people to make sense of their environment by providing certainty in how to interact with and think about others and developing a sense of self-identity through affiliation with various social groups (Ashforth and Mael, 1989). In organisations where people from diverse backgrounds accumulate, social sub-groups tend to form according to the similarity of various personal characteristics such as ethnicity, job role, and employment status (i.e., contractor versus permanent worker). Importantly, organisational research has shown that when the boundaries between these social groups are threatened (e.g., increasing the dissimilarity between groups or the number of members in one group compared to another), employees are less likely to trust one another, communicate frequently, or communicate efficiently. In the case of the 'out-group,' they are also more likely to disengage from their job tasks and their organisation (Rubino-Miner and Reed, 2010).

As contractors are typically smaller in number than permanent staff or are at least employed on unique terms (e.g., temporary status, different pay levels and benefits), differential treatment, whether perception or reality, is common. At the organisational level, contractors are often hired to perform the most dangerous or risky jobs within the confines of tightly defined employment contracts that emphasise the transactional nature of their work arrangement (Clarke, 2003). This approach reinforces group norms and stereotypes about contractors (e.g., 'not here for the long haul,' 'only in it for the money'), and may also contribute to less favourable perceptions of risk and organisational safety climate by the entire employee population encompassing contractors and permanent employees (Mearns, Flin, Gordon and Fleming, 1998).

A contractor's limited social position within an organisation can also reduce opportunities to communicate with other employees or participate in formal or informal safety training (Clarke, 2003). In addition, research has demonstrated that as compared to permanent workers, contractors are less committed to organisational goals and values (de Gilder, 2003) and less satisfied with their jobs (Biggs and Swailes, 2006)—key predictors of both work (Judge, Thoresen, Bono and Patton, 2001) and safety (Clarke, 2010; Christian, Bradley, Wallace and Burke, 2009) performance. In the case of contractor safety, low organisational commitment and job satisfaction may manifest as lack of involvement in safety activities, less knowledge of or interest in health and safety, less positive safety attitudes, and more emphasis on personal safety over that of co-workers (Burt, Sepie and McFadden, 2008; Clarke, 2003; Lipponen and Leskinen, 2006).

Overall, contractors' perceived and experienced differential treatment stems from implicit social dynamics that are present in everyday interactions between groups of people at work. Where differences between these groups are salient, such as clear discrepancies in employment status or the nature of job duties, differential treatment is likely to be amplified, which in turn impedes the development of a strong positive safety climate and reduces workers' ability to respond effectively in safety-critical situations.

Integrating contractors into an organisation's safety climate

Recognising everything described above reflects inherent risks to an organisation's safety climate when employing a contractor workforce, there are interventions that can target these risks and seek to mitigate their negative effects. Included below are recommendations for how to more effectively integrate contractors into the social structure of a client organisation with the goal of fostering a strong positive safety climate.

Contractor induction processes. The onboarding period for contractors is a critical juncture that significantly informs future client organisation-contractor relations. At this point, contractors have minimal information about the client organisation, so the 'data' they gather during this induction period have substantial influence on the contractors' impressions of the value the client organisation places on them and their safety. The socialisation phenomenon described above starts with the onboarding or induction process, so this is a prime target for intervention. The first step is to acknowledge that contractor onboarding should serve to not only provide contractors with site- and job-relevant knowledge so that they can perform their role effectively but also socialise the contractors into the organisation so that they can begin to integrate into established workgroups. The second step is to evaluate current induction processes to determine the extent to which both of these objectives are being accomplished. Where areas for improvement are identified, there may be several intervention opportunities such as involving more frontline employees (both contractor and permanent status) in induction processes, assigning a 'new hire buddy' to contractors to facilitate the socialisation process, creating formal or informal opportunities for senior management to introduce themselves to contractors, and sharing more information with contractors about the psychological elements of safety onsite, including the safety climate, rather than just technical information about pertinent safety regulations and requirements.

Safety leadership training. In most contractor arrangements, there is a line of supervision from the client organisation to the contractor. These client organisation supervisors serve a critical role in creating the social context that contractors work within. Put simply, if the supervisor does not treat contractors well does not attempt to integrate the contractors into the workgroup, no one will. Unfortunately, it is a misassumption that supervisors always possess the knowledge, skills, and motivation to effectively lead (not manage) their contractors. Frankly, it is unlikely that these supervisors were promoted into their position of authority because they were adept at inspiring and motivating a diverse team. It is also unlikely that these supervisors have received much 'soft-skills' training on how to create team safety cohesion or drive a strong positive safety climate within their workgroup. With this in mind, a potential intervention is to equip supervisors with these skills so that they can role model the effective integration of contractors into the workgroup. It should be noted that these types of skills would be relevant and important for supervisors to possess even if they do not formally supervise contractors onsite. As mentioned above, informal social dynamics are powerful within organisations; as such, supervisors can do much to lead towards respectful, fair, and safe treatment of contractors onsite regardless of formal reporting structures.

Cultural awareness training. Beyond safety leadership training, there is another training intervention opportunity to build a stronger more positive safety climate, in this case focussed on cultural awareness for the entire workforce. Cultural awareness training has long existed, with it most often being used in the context of white-collar professionals taking on expatriate roles in foreign countries. Today, cultural awareness training is relevant for many workforces, not just white-collar professionals on overseas assignments. For instance, many organisations in the heavy resource industry operate globally, or workers migrate from diverse cultures to work within the heavy industry. In both cases, cultural awareness training may provide employees with the skills to more effectively work within diverse teams, for the benefit of both safety and productivity. Whereas supervisors would particularly benefit from cultural awareness training, there is real value in providing this type of training to all workers, given evidence that co-worker interactions, including co-worker safety support and team safety cohesion are significant contributors to workgroup safety climate, and consequently team safety performance.

Joint client organisation–contractor safety initiatives. Along with new onboarding and training processes, there are also practice-related interventions that may be used to build stronger safety relations between contractors and client organisations. In particular, formal initiatives targeted at aligning safety across all organisational entities working at a site or on a project can be implemented. These might include contractor safety forums, joint safety committees, or safety recognition programs specifically designed to acknowledge effective safety communication between permanent workers and contractors. These types of initiatives are effective because they visibly and concretely provide evidence that management prioritises and values safety across the workforce, regardless of contractor or permanent employee status. As indicated above, programs designed and put into place by management can have a real and significant effect on employees' safety climate perceptions.

It should be noted that such an initiative has the potential to be helpful or hindering to the safety climate dependent upon how the initiative is executed. If contractors feel that management's effort is genuine and that they are being treated as partners (e.g., their input is being solicited and taken seriously during joint events), then this type of initiative can have powerful positive impact on safety climate. Indeed, some of the best contractor safety forums may even include presentations by certain contractors that have an exemplary safety performance records. Alternatively, more negative safety climate perceptions may develop if the initiative is treated as a way for the client organisation to chastise contractors for poor safety performance or tell them about how they should be doing things when it comes to safety. Overall, inviting contractors to participate in organisational safety initiatives and creating specific initiatives that are meant to be jointly run by the client organisation and contractors are effective ways to integrate contractors into an organisation's safety climate.

Moving forward towards a strong positive safety climate

The recommendations offered above are meant to stimulate new thinking about the ways that contractors can be better integrated into an organisation's social structure. Of course, there are many more ways to accomplish this goal, given an organisation's particular structure and practices as well as its current approaches to contractor safety management. All of these potential interventions are aimed at fostering a strong, positive safety climate across the entire workforce, with the goal of improving safety behaviour and ultimately safety outcomes. As organisations work towards this goal, it is imperative that they embrace the psychological and social element of contractor safety management as an excellent opportunity to improve contractor relations and workplace safety.

References

Allen, D. (2006). Do organizational socialization tactics influence newcomer embeddedness and turnover? *Journal of Management, 32*(2), 237–256.

Ashforth, B. (1985). Climate formation: Issues and extensions. *Academy of Management Review, 10*, 837–847.

Ashforth, B. and Mael, F. (1989). Social identity theory and the organization. *The Academy of Management Review, 14*(1), 20–39.

Beus, J., Bergman, M. and Payne, S. (2010). The influence of organizational tenure on safety climate strength: A first look. *Accident Analysis and Prevention, 42*(5), 1431–1437.

Biggs, D. and Swailes, S. (2006). Relations, commitment, and satisfaction in agency workers and permanent workers. *Employee Relations, 28*(2), 130–143.

Boyce, A., Ryan, A., Imus, A. and Morgeson, F. (2007). Temporary worker, permanent loser? A model of the stigmatization of temporary workers. *Journal of Management, 33*(1), 5–29.

Burke, M., Chan-Serafin, S., Salvador, R., Smith, A. and Sarpy, S. (2008). The role of national culture and organizational climate in safety training effectiveness. *European Journal of Work and Organizational Psychology, 17*(1), 133–152.

Burt, C., Sepie, B. and McFadden, G. (2008). The development of a considerate and responsible safety attitude in work teams. *Safety Science, 46*(1), 79–91.

Clarke, S. (2003). The contemporary workforce: Implications for organisational safety culture. *Personnel Review, 32*(1), 40–57.

Clarke, S. (2006). The relationship between safety climate and safety performance: A meta-analytic review. *Journal of Occupational Health Psychology, 11*(4), 315–327.

Clarke, S. (2010). An integrative model of safety climate: Linking psychological climate and work attitudes to individual safety outcomes using meta-analysis. *Journal of Occupational and Organizational Psychology, 83*(3), 553–578.

Christian, M., Bradley, J., Wallace, J. and Burke, M. (2009). Workplace safety: A meta-analysis of the roles of person and situation factors. *Journal of Applied Psychology, 94*(5), 1103–1127.

de Gilder, D. (2003). Commitment, trust, and work behaviour: The case of contingent workers. *Personnel Review, 32*(5), 588–604.

Hofstede, G. (1993). Cultural constraints in management theories. *Academy of Management Perspectives, 7*(1), 81–94.

Judge, T., Thoresen, C., Bono, J., and Patton, G. (2001). The job satisfaction-job performance relationship: A qualitative and quantitative review. *Psychological Bulletin*, 127(3), 378–407.

Kouabenan, D. (2009). Role of beliefs in accident and risk analysis and prevention. *Safety Science, 47*(6), 767–776.

Lindell, M. and Brandt, C. (2000). Climate quality and climate consensus as mediators of the relationship between organizational antecedents and outcomes. *Journal of Applied Psychology, 85*(3), 331–348.

Lingard, H., Cooke, T. and Blismas, N. (2010). Safety climate in conditions of construction subcontracting: A multi-level analysis. *Construction Management and Economics, 28*(8), 813–825.

Lipponen, J. and Jukka, J. (2006). Conditions of contact, common in-group identity, and in-group bias among contingent workers. *Journal of Social Psychology, 146*(6), 671–684.

Lipponen, J. and Leskinen, J. (2006). Condition of contract common in-group indentity, and in-group bias towards contigent workers. *The Journal of Social Psychology*, 146(6), 671–684.

Mearns, K., Flin, R., Gordon, R. and Fleming, M. (1998). Measuring safety climate on offshore installations. *Work and Stress, 12*(3), 238–254.

Park, Y. and Butler, R. (2001). The safety costs of contingent work: Evidence from Minnesota. *Journal of Labor Research, 24*(4), 831–849.

Renn, O. and Rohrmann, B. (2000). *Cross-Cultural Risk Perception: A Survey of Empirical Studies*. Kluwer Academic Publishers: Dordrecht, Netherlands.

Rogers, J. (1995). Just a temp: Experience and structure of alienation in temporary clinical employment. *Work and Occupations*, 22(1), 137–166.

Rousseau, D. (1988). The construction of climate in organization research. In C. Cooper and I. Robertson (Eds.). *International Review of Industrial and Organizational Psychology, 3*, Wiley: Chichester.

Rubino-Miner, K. and Reed, W. (2010). Testing a moderated mediational model of workgroup incivility: The roles of organizational trust and group regard. *Journal of Applied Social Psychology, 40*(12), 3148–3168.

Schneider, B., Goldstein, H. and Smith, D. (1995). The ASA framework: An update. *Personnel Psychology, 48*(4), 747–773.

Schneider, B. and Reichers, A. (1983). On the etiology of climates. *Personnel Psychology, 36*(1), 19–39.

Schutle, M., Ostroff, C., Shmulyian, S. and Kinicki, A. (2009). Organizational climate configurations: Relationships to collective attitudes, customer satisfaction, and financial performance. *Journal of Applied Psychology, 94*(3), 618–634.

Starren, A., Hornikx, J. and Luijters, K. (2013). Occupational safety in multicultural teams and organizations: A research agenda. *Safety Science, 52*(1), 43–49.

Vickers, I., Baldock, R., Smallbone, D., James, P. and Ekanem, I. (2003). Cultural influences on health and safety attitudes and behaviour in small businesses. *Health and Safety Executive Research Reports*. HSE Books: London.

Yagil, D. and Lunia, G. (2010). Friends in need: the protective effect of social relationships under low-safety. *Ultimate Group Organization Management*, 35(6), 727–750.

Zohar, D. (1980). Safety climate in industrial organizFrations: Theoretical and applied implications. *Journal of Applied Psychology, 65*(1), 96–102.

Zohar, D. (2010). Thirty years of safety climate research: Reflections and future directions. *Accident Analysis and Prevention, 42*(5), 1517–1522.

Zohar, D. and Luria, G. (2004). Climate as a socio-cognitive construction of supervisory safety practices: Scripts as proxy of behavioural patterns. *Journal of Applied Psychology, 89*(2), 322–333.

Zohar, D. and Tenne-Gazit, O. (2008). Transformational leadership and group interaction as climate antecedents: A social network analysis. *Journal of Applied Psychology, 93*(4), 744–757.

Index

H

I

J

K

L

M